LESS HEAT
MORE LIGHT

A GUIDED TOUR OF
WEATHER, CLIMATE, AND
CLIMATE CHANGE

JOHN D. ABER

Yale

UNIVERSITY PRESS

New Haven and London

Published with assistance from the foundation established in memory of Philip Hamilton McMillan of the Class of 1894, Yale College.

Yale University Press books may be purchased in quantity for educational, business, or promotional use. For information, please e-mail sales.press@yale.edu (U.S. office) or sales@yaleup.co.uk (U.K. office).

Printed in the United States of America.

Library of Congress Control Number: 2022913375
ISBN 978-0-300-25943-8 (hardcover : alk. paper)

A catalogue record for this book is available from the British Library.

This paper meets the requirements of ANSI/NISO Z39.48-1992 (Permanence of Paper).

10 9 8 7 6 5 4 3 2 1

Dedicated to Kyla, Nolan, and John
You are the future

Contents

CONTENTS

ROOM THREE

EARTH SYSTEM SCIENCE AS A PROCESS:

HOW IT WORKS, AND WORKS FOR US

ROOM FOUR

CLIMATE CHANGE: THE DATA, THE MODELS,

AND OUR FUTURE

Preface

I have always been a climate geek. This may sound strange for a boy who grew up in a place some would say has no climate. As a child in Los Angeles, I was always wishing for more rain, more storms, more weather. The one time it did snow at our house, I was outside in a flash making a snowball that was stored in the freezer until it evaporated (sublimation being the technical term). In an area averaging less than 14 inches (35 cm) of rain a year, and often receiving much less, I put a rain gauge in the backyard and kept wishing for something besides spiders to fall into it.

Had there been a meteorological major in the college I attended, that would have been my choice, but liberal arts colleges aren't known for their applied fields of study. Still, as my career in environmental science has unfolded, climate and weather have always been a part of my research and teaching and a never-ending source of fascination and intellectual challenge. Every semester I have taught classes that included updates in how we measure, study, understand, and predict both weather (with a time frame of hours to days) and climate (years to centuries).

In the twenty-first century, immersion in weather information has become commonplace. Is there any other topic in daily conversation for which we have instant access to such amazingly detailed images, driven by a process that ingests such huge amounts of data instantaneously? I suppose the stock markets also provide instant access to large amounts of timely information, but the weather community also makes predictions days into the future. With stocks, those predictions are left to the pundits.

How this access to weather data has changed! As a ten-year-old with such a strong interest in weather, I received from my parents a subscription to the daily weather maps printed by the U.S. Weather Service. These arrived by mail three days after the fact! As a young professor at the University of Virginia, my department was privileged to have access to huge, high-resolution hard-copy maps of current conditions from the Weather Service that were printed every few hours. When the Weather Service started broadcasting a live show with current information and maps, that was truly compelling, and educational as well. It is hardly worth mentioning how primitive all of this seems now compared with the instantaneous radar and visible images available in real time on every Web-based device from governmental and commercial sources.

Yes, we now have unprecedented access to timely information from all parts of the complex, three-dimensional climate system, and a thorough understanding of the basic physics that drive weather and climate. And yet our ability to predict weather more than a few days out runs into the wall called chaos (the mathematical variety). This is just one fascinating aspect of the weather story I explore in this book.

But what about climate change? If we can't predict the weather in the near future, how can we possibly predict climate decades in advance? Although the proposition sounds logical, weather and climate, while obeying the same relatively simple physical laws, are constrained by different variables at different scales—an example of what is called systems analysis. The same system can be controlled and best predicted or explained by different factors at different time and space scales. Sounds illogical, but that is another part of the weather and climate story I find compelling and try to present.

There is not a lot of new science here. In fact, one of the central themes of this book is that we have known most of the important facts about weather and climate for a long time. The analogy of a museum with thematic rooms comes to mind, and I propose to be your guide through four of these rooms focused on historical context, the basics of weather and climate, how environmental research is carried out and applied, and how this all relates to our changing climate system. A guide can add value to the works on display by enriching the stories apparent in the works themselves. My role here is to dovetail different parts of the weather/climate story into a single coherent canvas.

More specifically, I hope to convey three central ideas.

The first is that the basic science behind both weather and climate is well known, and has been for decades. Although there are remaining mysteries, especially around the role of clouds, the dynamic future of ice, and the impacts of climate oscillations such as El Niño, there is more than enough known to understand the basic patterns and processes, and to predict where we are headed. We also understand how we can predict changes in climate over the coming decades even though the accuracy of weather forecasts fades rapidly more than a week or two ahead. Beyond the cold, hard facts, I also try to communicate some of the fascination I feel watching weather and climate unfold with some basic understanding of why and how we see what we see.

The second central idea is that there is pattern and structure to how we increase our understanding of the workings of the Earth as a single, integrated system through a unique field of research now called Earth system science. The pattern is similar whether answering the big, basic questions about the history and workings of our planet, or how those answers can be applied to environmental concerns.

There really are few if any "aha" or epiphanic moments in science at this level. Major questions are resolved through a long period of give-and-take, measurement, and synthesis. We will see just how long the time line can be between crazy idea, initial dismissal, extended research by many, and final resolution. I pay some special attention to the role of computer models, as they have become integral to the understanding and presentation of science, and have been a focus of my own career in science from the very beginning.

Finally, all of this background and context about weather and climate, as well as how science happens, are applied to the question of climate change. The point here is that we know enough to understand where we are headed, and have known that for some time. My hope is that a guided tour of the science as presented in this book can help foster a public discussion of climate change based on a firmer understanding of the science, and support a climate debate generating less heat and more light!

Acknowledgments

So many people have contributed to the crafting of this book. A first thank-you goes to all the students in my classes over the years who have taught me as much as I have taught them. The give-and-take in the classroom is a constant source of inspiration and challenge. Putting together an Excel lab for one of those classes using the Second Arrhenius Equation is actually what planted the seed for this guided tour of weather and climate. Special thanks to Stephanie Petrovick, who provided a critical student's perspective on the work.

Allison Leach has been deeply involved in this project from the beginning and has provided a colleague's timely and insightful reviews, as well as guidance and technical support on text and figures.

Several colleagues have read, reviewed, and offered honest commentary on the text as it has evolved. David Foster helped frame the direction of the book and has been a long-term source of information and understanding of environmental systems. Jerry Melillo has been inspirational in more ways than he may know. John Pastor has read the manuscript from start to finish and helped to refine and clarify many points. Scott Ollinger has been a long-term colleague and was an important catalyst for this project, along with Alix Contosta, by sponsoring a special Earth Day webinar. Larry Mayer provided an important perspective on the role of science and policy. Gus Speth and Charles Canham provided detailed and helpful reviews as well. A special thank-you to Bill Peterjohn for a critical look into the origin of the formulation that I have called the Second Arrhenius Equation.

Darrell Ford and Jim Lewis took the time to read the entire manuscript and provide the additional perspective of educated readers from outside the sciences. Caitlin Aber contributed a historian's perspective and a very careful reading of the entire text.

While working on this book, I was also posting essays on similar topics at Substack.com. Many colleagues and friends commented on these essays, and our conversations helped sharpen both the content and the writing of those essays and this book. My thanks to Susan Antler, Liz Burakowski, Alix Contosta, Mark England, Thomas Frederikse, Joe Hendricks, Nicholas Herold, Richard Houghton, Matthew Huber, Brian Jerose, Norman Loeb, Chuck McClaugherty, Bill Munger, Knute Nadelhoffer, William Putman, and Jonathan Thompson.

The idea of simpler explanations of climate change was part of a webinar at a recent class reunion, and thoughts and comments shared by Lupi Robinson, Frances Beinecke, Walter Minkeski, and Meridith Wright provided valuable direction.

Many thanks to Jean Thomson Black at Yale University Press, who has provided support and guidance throughout the project. Her professional experience has been crucial in expanding the scope of reviews of the manuscript and in shaping the title and the structure of the book. It has been a pleasure to work with her again. Thanks also to Elizabeth Sylvia and Hannah Alms for providing expertise on text and figure logistics. Otto Bohlmann assured consistency and accuracy of the text, and Joyce Ippolito found many ways to improve the structure of the work and provided many insights on best ways to tell this story.

And finally, thanks to all the family, especially Lynn, Patrick, Colleen, Caitlin, Jin, Kenny, Alexandra, and of course, Kyla, Nolan, and John, for listening to weather and climate stories for so many years!

A Note on Units

Most of the countries in the world, and all of the scientific organizations and publications, use metric or SI units to express weight (kilograms), distance (meters), temperature (degrees Celsius), and other variables. The United States is essentially alone in using imperial units (pounds, miles, degrees Fahrenheit) for both public and private communications. U.S. agencies like the Weather Service, NOAA (National Oceanic and Atmospheric Administration), and NASA carry out their business in metric units but often report publicly in imperial units. I use metric units throughout this book but also give imperial units in a few places where the numbers will be more familiar to many readers or where graphics embedded in the text use that system. For temperature, F stands for degrees Fahrenheit and C for degrees Celsius.

ROOM ONE
Weather and Climate
SIMPLE OR COMPLEX?

✧

1
Was Svante Right?

In what year did an early Nobel Prize–winning scientist make a first and surprisingly accurate calculation of how increasing carbon dioxide in the atmosphere would increase global temperatures? Who was that scientist?

Figure 1.1. Svante Arrhenius in 1909.

I have used these questions at the beginning of presentations in class or to general audiences as a way to break the ice and get the conversation started. The answers are almost always a surprise. The when is 1896, and the who is Svante Arrhenius (figure 1.1).

Arrhenius will be a touchstone throughout this book as his story captures many of the themes I present.

The first is that we know a lot about weather and climate, and have known much of it for a very long time. In a book written for general audiences in 1908, Arrhenius outlined a surprising number of the basic processes affecting carbon dioxide in the atmosphere and its impact on climate, including the role of burning fossil fuels.

A second theme is the value of stepping outside traditional academic boundaries to address the big questions. Arrhenius's intellectual journey was a nontraditional one, even though he did win the biggest of all scientific prizes—the Nobel. The award was given for his work in physical chemistry. The interaction of carbon dioxide and climate was an intriguing sideline for him that occupied most of a year of grueling hand calculations. We will see that many early climate calculators pursued this field as a sideline to their main careers.

A third is the role of unstructured academic environments. That Arrhenius undertook his groundbreaking calculations on carbon dioxide and temperature at all can be credited in part to the innovative academic environment of the Stockholm Högskola, where he was offered his first academic position following years of traveling from lab to lab across Europe.

A final theme is the importance of working with a stimulating set of colleagues from a wide range of backgrounds. When Arrhenius finally landed at the Högskola, he convened a rich assembly of inquiring minds under the aegis of the Stockholm Physics Society. This group stimulated Arrhenius's attack on the question of carbon dioxide and climate, as well as groundbreaking work by others in meteorology, biogeochemisty, and global carbon cycling.

Arrhenius's academic history, more than a century old, captures many characteristics of the modern study of climate change and its impacts. Much of the work is done in interdisciplinary labs and institutions that foster intense intellectual exchange and the constant give-and-take that makes the work of science both stimulating and enjoyable. Earth system science is one modern name given to the study of the Earth as an integrated system, and I feel it is not too far-fetched to call Arrhenius one of the founders of that field, in addition to his Nobel-winning work in physics and chemistry.

So, more than a century later, Svante Arrhenius has inspired the creation of this book. In search of simpler ways to tell the climate change story, I read more deeply into the somewhat dismissive statements usually made about his early work on carbon dioxide and climate, and discovered a brilliant and wide-ranging intellect who was also dedicated to sharing his own very evident joy in the process and outcome of science with a broad audience.

For all these reasons, Arrhenius appears with some regularity throughout these pages. Given that, I think that retelling his story here in a little more detail can serve as a valuable prelude to our guided tour of weather and climate. The only major biography of Arrhenius written in English is Elisabeth Crawford's *Arrhenius: From Ionic Theory to the Greenhouse Effect*. This brief summary of his life draws heavily on that source.

Svante Arrhenius was born to a Swedish farming family in 1859. His father was not the first son in his generation and so could not inherit the farm. As a result, the family moved to a small town near Uppsala, home to the oldest university in Sweden. One of Svante's uncles had earned a Ph.D. at Uppsala, had established a successful career there, and is associated with the founding of its Agricultural Institute.

Svante's uncle helped his father obtain the position of rent collector for the university, a position in which he also served for a local noble family. Svante's father was a conscientious manager who was able to leave his children a "small capital bequest."

Svante was tutored at home, attended local schools, and was admitted to the university at Uppsala in 1876. He was an above-average student with a facility with numbers and mathematics, always able to carry out very complicated calculations in his head.

As an undergraduate, Arrhenius studied physics and chemistry, but the humanistic ideals that guided the undergraduate curriculum at Uppsala also required studies in Latin, history, geology, and botany. He completed his undergraduate work in only four semesters and began graduate studies at Uppsala, working with and between the programs in physics and chemistry.

At this time, Uppsala, and most European universities, maintained a strict hierarchical structure where each discipline and department was managed by a single professor who exercised complete control over personnel, budget, and curriculum. Arrhenius's interest in both physics and chemistry displays creativity and a tendency to think across traditional disciplines that would mark his later successes but was unlikely to be appreciated by the professors in each department. One said he had "a brilliant mind" but thought him lazy and clumsy, concluding, "No doubt he has ability, if only he could show it."

But revolution, at least of the academic sort, was brewing in Uppsala. Student unrest with the primitive laboratory conditions and the antiquated curriculum in the sciences led to calls for reform, to which entrenched professors offered stern resistance. Students looking for better opportunities, including Arrhenius, turned to the Institute of Physics at the Academy of Sciences in Stockholm for better facilities and a more open research environment.

Another venue to be of special value to Arrhenius also emerged from this period of controversy and growth. In 1865, the Stockholm City Council set up a fund and a committee composed of nine prominent citizens to establish an institution of higher education in the capital.

As a result, the Stockholm University College (Högskola in Swedish) was chartered in 1877 and began operation in 1878 with a series of lectures on the natural sciences, open to general audiences. For quite some time it was essentially a private, non-degree-granting institution and was notable, among other things, for the appointment of Sofia Kovalevskaya in 1889 as chair in the mathematics department, making her only the third female professor in Europe. It was also an "open" university, with unstructured curricula and an absence of exams and grades.

Arrhenius spent a good deal of time in Stockholm from 1881 to 1883 while also pursuing a Ph.D. in chemistry through the university at Uppsala. His absence from Uppsala allowed him to focus on a research question that his primary professor in Uppsala had explicitly told him not to pursue. Arrhenius did not see chemistry and physics as distinct fields. The freer environment in Stockholm, at both the Högskola and the Academy of Sciences, allowed him to pursue the work that would eventually lead to the founding of the new combined field of physical chemistry, and to the Nobel Prize.

Arrhenius had hoped to earn his Ph.D. through the Högskola, feeling that the ability to grant such degrees would be approved in short order. That did not happen. The Stockholm Högskola, in its gradual transition into what is now Stockholm University, did not become a degree-granting institution until 1904, and not until 1960 did it achieve university status, becoming Sweden's fourth state university.

Arrhenius produced an unusual thesis in two parts. The first part was a more traditional presentation of research findings on the conductivity of electrolytes in solutions. The second part was apparently something of a theoretical ramble, including fifty-six hypotheses or statements.

This approach did not sit well with his official professors in Uppsala, and the outcome was not favorable. Rather than being awarded the highest degree possible for his work, he was given instead a fourth-class Ph.D., which effectively eliminated any chance of his continuing on at the next level in the academic hierarchy at Uppsala. Beyond that, his two major professors refused to recognize him during the very formalized process of celebrating each new Ph.D. graduate, instead donning their coats and exiting the ceremony early and in haste. One colleague present at the disputation, or thesis defense, recorded these events decades later and said that Arrhenius's professors had decided to sacrifice him as they considered theories and hypotheses worthless.

The irony here is that the initial work that revolutionized both chemistry and physics by joining them in a new field of physical chemistry was criticized for venturing too far from the disciplinary norms of the day. His theory of the forces controlling the dissociation of electrolytes led to the Nobel Prize in 1903, and his derivation of the Arrhenius Equation for the temperature dependence of chemical reactions remains a standard part of introductory chemistry classes. Both combine physical and chemical concepts.

But in 1884, Svante Arrhenius was a young scientist at home in neither of the traditional disciplines of chemistry and physics, with a dismissively low grade on his thesis, and no apparent future in his home country. He was described, though, as an energetic, affable, argumentative, confident young man. For Arrhenius, scientific argument was fun, and not necessarily cause for personal animosities.

What I find so compelling about Arrhenius's story is that he was not cowed by his experience at Uppsala. Displaying an unusual level of confidence for one with no notable success in a traditional field, he presented his thesis and associated papers to well-known scientists across Europe. There he found appreciation and support, and the ability to share ideas with many colleagues across different fields of study and different institutions. He spent six years traveling among laboratories headed by leading scientists in Latvia, Germany, Austria, and Holland.

It may also be both ironic and indicative that Arrhenius never did achieve the coveted status of professor at a major research university, through which he could have commanded both research and teaching agendas. In fact, at one point he decided against accepting a second-tier professorship in chemistry in Germany, saying that he really considered himself a physicist but also citing the "superhuman diligence" required of professors in that country. He questioned whether he could live up to the demands that would be put on him, but perhaps there was also a desire to avoid the highly structured and narrowly focused environment in which he would have had to work.

Arrhenius's academic wandering finally ended when he secured a position, not at Uppsala, but at the Stockholm Högskola in 1891. He was thirty-two years old and seven years past his Ph.D. A private, non-degree-granting institution was not the natural home for a soon to be Nobel-winning scientist, but the free-wheeling structure of this institution must have allowed and supported his wide-ranging intellect. It also provided him with his first actual laboratory for his own work.

His first position at the Högskola was as a lecturer—primarily a teaching position—but his recognition outside Sweden led to a stream of aspiring students coming to Stockholm to work with him. He was eventually promoted to the position of professor of physics in the Högskola in 1895 over much opposition, but this was not the same as a position with that title at a major university like Uppsala. In 1896 he became rector of the Högskola, its most senior academic official, showing perhaps an interest in supporting and managing this unusual academic institution. In 1903 he was elected to the Royal Swedish Academy of Sciences—again over strong opposition.

But his tenure at the Högskola was also not a long one, by university standards of the time. In 1900 he became involved in the establishment of the Nobel Prizes and was a member of judging committees for many years. He won the prize in chemistry in 1903 and in 1905 was appointed rector of the newly established Nobel Institute for Physical Research, located in Stockholm. There he remained until his retirement in 1927. He apparently preferred this nontraditional academic home to the possibility of traditional professorships at Swedish universities. At one point he described the Högskola as a place where there were people he could talk to. Perhaps traditional universities would have been too isolating and limiting.

One of Arrhenius's first acts in his new academic home in the Högskola in 1891 was to establish the Stockholm Physics Society. This was a fluid organization structured around biweekly meetings of leading scientists across all major disciplines, as well as others in Stockholm society who would be interested in broad-ranging, free-wheeling discussions of current and nontraditional scientific topics.

This new organization was in part a reflection of its time and place in Swedish history, for the 1890s was a time of elevated national pride, of creative living and an adventurous spirit. The Högskola can be seen as one expression of this free-thinking approach, but Swedish explorers were also engaged in several dangerous and novel incursions into the Arctic world, with ambitions of reaching the north pole. It was a time of possibilities, of escape from limited ambitions and narrow thinking.

Arrhenius was a man of his times in this regard. Even before receiving the Nobel Prize, he had left behind laboratory-based research in physical chemistry, perhaps because he felt the major questions had been answered and were no longer interesting or challenging. Rather than pursuing second-order questions or settling into a comfortable professorship and spending his later years resting on the laurels of his Nobel Prize, he remained incurably curious about science in the broadest sense and recognized no boundaries among scientific topics.

So the Stockholm Physics Society encouraged and supported conversations among people with wide-ranging areas of experience and expertise that would have been hard to convene within the highly structured setting of a major university of that time.

One all-inclusive topic Arrhenius pursued with his colleagues in the society was what he termed cosmic physics. Plowing through his language on this in different sources, I realized that he was basically describing what we now call Earth system science—treating the whole Earth as an interactive system encompassing processes in chemistry, physics, biology, and more. It may not be too farfetched to call him a founder of Earth system science in the modern sense of a rigorous scientific field with hypotheses based on physical principles, as well as the founder of physical chemistry.

But for Arrhenius, cosmic physics went well beyond the Earth and dealt with other planets, the sun, the solar system, and just about everything else. As one definition of genius has it, he saw similarities where others would see differences.

It is the intellectual hotbed of the Stockholm Physics Society that brings us to Arrhenius's work on carbon dioxide and climate. Imagine one of the biweekly meetings of the society. Some in the room had been part of expeditions to the Arctic; the affinity for stories about ice was a natural. The idea that ice ages had covered much of Europe, including Sweden, with glaciers more than a kilometer thick had come to be generally accepted in scientific circles, but the causes of both the onset and the retreat of such massive amounts of ice were unknown. Curiosity about ice ages and concern for the possible return of the massive ice sheets lent some social incentive to answering the question, such that in 1894 a prize was offered through a scientific organization for the best scientific explanation.

Also present at this imagined meeting might be the four scientists in addition to Arrhenius who provided much of the intellectual vitality and broad perspective of the society. Otto Pettersson was a chemist turned oceanographer with speculative interests in the interactions between the oceans and the atmosphere. Nils Ekholm was a meteorologist engaged in organizing Sweden's national weather service. Arvid Högbom was a student of Arrhenius's who turned from chemistry to geology. He was really a geochemist before the term was coined and an early expert on what we now call the global carbon cycle, including the factors controlling the concentrations of that element in the atmosphere. Vilhelm Bjerknes was a young mathematical physicist who, during his time at the Högskola, established the physical basis of the role of pressure differences in the atmosphere in driving patterns of circulation—and weather. It has been said that Bjerknes's theories of atmospheric circulation, the foundation of modern meteorology, might not have been developed without the open yet rigorous environment of the Högskola and the Stockholm Physics Society.

At our imagined meeting of the society, let's say that a presentation is made about ice ages, and discussion ensues about the physical processes underlying this major dynamic in the Earth system. Arrhenius, who is also aware of research carried out decades earlier on the heat-trapping property of carbon dioxide and its role in altering the temperature of the atmosphere, might speculate that changes in the concentration of this gas were responsible for ice ages. Högbom could describe the long-term balance of carbon dioxide in the atmosphere due to geologic processes. Could these processes act swiftly enough to drive the ice ages? Pettersson could discuss the role that oceans play as a "sink" for carbon dioxide as the gas dissolves in seawater. Ekholm and Bjerknes could add the perspective of meteorology.

It would have been fascinating to be a fly on the wall for this kind of discussion, one that would have been very unlikely to have occurred within the hierarchical confines of a traditional university, and one perhaps unlikely ever to find a permanent home within that structure. The fact that cosmic physics as an intellectual enterprise disappeared in the early 1900s, having never found a home in a university department, emphasizes both the broadly interdisciplinary nature of the topic and the unique and valuable intellectual environment of the Stockholm Physics Society.

It was his curiosity about ice ages, combined with the established knowledge about the heat-trapping properties of carbon dioxide and some preliminary data on the cycling of carbon (all topics for discussion in the society), that set Arrhenius on the arduous set of calculations that led to the derivation of what I will call the Second Arrhenius Equation. As early as 1894, in response to that competition to explain the occurrence of ice ages, he presented a paper describing the possible role of reductions in the carbon dioxide content of the atmosphere in triggering an ice age.

The mid-1890s was a tumultuous time for Arrhenius. His short-lived marriage to a brilliant and independent-minded scientist, another reflection of Sweden in the 1890s, was dissolving. He was embroiled in a major controversy about his possible appointment as professor in physics at the Högskola, and many of his existing colleagues at other institutions were confused by the type of position he held, the institution in which he held it, and his jump from physical chemistry to something called cosmic physics. The Högskola itself was embroiled in controversy regarding its unstructured and free approach to education.

Perhaps against this background, and with the stimulation provided by Arrhenius's Stockholm Physics Society colleagues, a deep dive into the topic of carbon dioxide and climate seemed appealing.

Arrhenius cites the extraordinary interest in the topic expressed by his society colleagues as important to him in deciding to undertake the massive calculations relating carbon dioxide and climate. He was to spend a full year carrying out those calculations on what he called at the end "a trifling matter."

And yet this trifling matter produced what could be termed one of the first global climate models. How did this model work? Central to Arrhenius's calculations was the established physics of the absorption of infrared (or "heat") radiation by both water vapor and carbon dioxide.

Carbon dioxide is now known to be the most important of human-derived greenhouse gases driving climate change. How long have we known of the ability of this gas to absorb infrared or "heat" radiation and to drive the greenhouse effect? This is another question I like to use in presentations and in class.

The generally surprising answer is more than 160 years, or since John Tyndall first posited in 1861 that changes in the concentrations of water vapor and carbon dioxide in the atmosphere "may have produced all the mutations of climate which the researches of geologists reveal."

Arrhenius not only knew about the process of absorbing heat radiation, he also had data on the relationship between atmospheric concentration and the quantity of radiant energy absorbed—essentially how the strength of the greenhouse effect varied with the amount of carbon dioxide in the atmosphere.

A characteristic of Arrhenius's work on this topic, which caused some undervaluation of the results by traditional colleagues, was that very little of what came to be important in his model was based on his own research. In this case, and throughout his research, he always gave full credit to the researchers who had developed the theories and equations that he used. In this sense Arrhenius also foreshadowed the role that mathematical modelers currently play in synthesizing the current state of understanding and predicting both weather and climate.

Although a first approximation of the effect of carbon dioxide on climate could be derived by calculating the direct effect of this gas, Arrhenius was aware, probably through conversations with the meteorologists in the society, that a warmer atmosphere could hold more water vapor (and conversely a cooler atmosphere, less water vapor). As Tyndall had known, water vapor is the most powerful of all greenhouse gases, so a more complete analysis of the impact of carbon dioxide must include changes in water vapor content as well. In either modeling or systems analysis, this is called a feedback, where one direct effect causes a second result that can either augment or reduce the impact of the initial effect. Arrhenius calculated the impact of decreased (or increased) carbon dioxide on water vapor content and the effect of that change in vapor on temperature, capturing this feedback.

From a year of calculation, Arrhenius concluded that about a 40 percent decline in carbon dioxide in the atmosphere could have initiated an ice age.

Our current understanding is that ice ages are initiated by subtle changes in the Earth's orbit and angle relative to the sun, reinforced by induced feedbacks in the Earth system, including changes in carbon dioxide, but one of the strengths of the general result Arrhenius produced is that it can also be used to predict increases in temperature in response to increases in carbon dioxide. This is what has placed Arrhenius in the pantheon of founders of climate change research.

I return to the details of Arrhenius's predictions and a simple equation summarizing his findings in chapter 11, and then describe how well that equation performs using twenty-first data. For now, let's complete this profile of Arrhenius by featuring his dedication to presenting science to general audiences, and how he framed his carbon dioxide work in that setting.

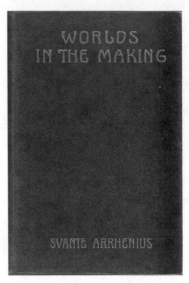

Figure 1.2. The English translation of Arrhenius's book on cosmic physics intended for a broad audience.

His master work summarizing cosmic physics for general audiences was a volume entitled, in English, *Worlds in the Making*, published in 1908 (figure 1.2). In this book he treats topics as diverse as the origin and fate of the sun, the possible origin of life on Earth, and geological phenomena like earthquakes and volcanoes, all as part of a general view of the universe. It won't be surprising that many of his speculations were not supported by later research, but the section on the heat budget of the Earth, one that summarizes that year of intense thought and calculation, contains a surprising amount of knowledge we now accept as part of Earth system science.

I was stunned the first time I encountered this list of statements. Remember that they were penned in 1908:

- "The temperature of the Earth's surface . . . is conditioned by the properties of the atmosphere." This statement is definitely true, and it was known before Arrhenius's time.

- "Their theory has been styled the hot-house theory, because they thought the atmosphere acted after the manner of the glass panes of hot-houses." This analogy is imperfect, as many have said, but the terminology, changed to greenhouse effect, persists.
- "Any doubling of the percentage of carbon dioxide in the air would raise the temperature of the Earth's surface by 4°." I test this prediction in chapter 11 and compare it with recent rigorous assessments. It is not far off.
- "The sea, by absorbing carbonic acid [another term for carbon dioxide], acts as a regulator of huge capacity, which takes up about five-sixths of the produced carbonic acid." The actual value is closer to 50 percent, but this is still an astonishingly early recognition of a major process in the Earth system and its influence on climate.
- "The enormous combustion of coal by our industrial establishments suffices to increase the percentage of carbon dioxide in the air to a perceptible degree." Arrhenius did not foresee the tremendous increase in the combustion of fossil fuels in the coming years and so predicted it would take centuries for this to have a major effect on climate.
- "By the influence of the increasing percentage of carbonic acid in the atmosphere, we may hope to enjoy ages with more equable and better climates, especially as regards the colder regions of the Earth, ages when the Earth will bring forth much more abundant crops than at present, for the benefit of rapidly propagating mankind." This passage reflects the concern over the relatively recently accepted occurrence of major ice ages, and the potential for a new ice age to obliterate much of Europe.

Arrhenius's writings and detailed calculations also included other aspects of the global climate system that have been verified by later research. These include:

- Greater warming at the poles, based on differences in "nebulosity" (humidity, clouds), angle of solar input, and the reflection of sunlight by snow and ice. Greater warming in the Arctic is one of the best documented effects of global climate change.
- Feedbacks with ice loss or gain and decreased or increased reflection of sunlight.
- Increased carbon dioxide uptake by plants at higher concentrations of this gas.

In addition to his prediction of a 4 degree increase in global temperature with a doubling of carbon dioxide, Arrhenius predicted an 8 degree increase with a quadrupling of carbon dioxide and decreases of 4 degrees and 8 degrees for reductions of 50 percent and 75 percent in this greenhouse gas. When graphed, this gives a nonlinear relationship that has been summarized for application in an era of increasing carbon dioxide concentrations using this equation:

$$\Delta F = \alpha * \ln (C/Co)$$

I won't decipher this equation here, or discuss the details of its origin, but in chapter 11 I offer a surprising test of its accuracy. For now, I will refer to this as the Second Arrhenius Equation, so named to honor his yearlong calculational marathon and first quantitative prediction of the effect of carbon dioxide on temperature, and in deference to the well-known (first) Arrhenius Equation relating to the rate of chemical reactions.

The Arrhenius story begins to tell us just how long the basics of the interactions of greenhouse gases and temperature have been known. His experience also demonstrates the value of the stimulation provided by a richly interdisciplinary and relatively unstructured academic and intellectual environment in catalyzing major leaps in our understanding of complex systems like the climate of the Earth.

But can the explanation and prediction of future climates really be captured and conveyed in a relationship as simple as the Second Arrhenius Equation? What about the hugely complex computer models that generate the trends and maps usually presented as descriptors of our weather and climate future?

Perhaps a medieval philosopher can help us dissect this apparent contradiction.

2

Occam's Razor and the Case for
Simpler Explanations

What might a thirteenth- to fourteenth-century philosopher and theologian be able to offer to contemporary Earth system science and our discussion of climate change? Perhaps not a lot when it comes to predicting tomorrow's weather, but there may be some wisdom here in conveying the basics of the climate system to a broader audience.

William of Occam (1287–1347) was an original and revolutionary thinker who confronted both the philosophical and theological barriers of his time, even accusing a sitting pope of heresy! One target of William's sharply critical mind was the elaboration of multiple, unrelated, and unsupportable explanations for physical phenomena (think crystal spheres to describe the movement of planets in the sky). While he wrote in Latin, one of the most famous translations of his approach is:

Entities are not to be multiplied without necessity.

A simple and powerful statement; explain a set of observations with as few parameters or entities as possible.

This concept has come to be called Occam's razor. In a recent book, JohnJoe McFadden credits Occam's influence and the resulting drive for simpler explanations with leading to many of the fundamental breakthroughs in physics.

History enjoys giving a single name to important concepts, but, as in most such cases, William's thoughts result from and reflect the intellectual environment of his era. And not surprisingly, there was at the same time an opposing school of thought—the anti-razors—and there have been positive and negative responses to the concept ever since—proving its value as a stimulant for analytical thinking.

Chapter 1 included a simple formulation of the impact of increased carbon dioxide on global temperature derived from Arrhenius's yearlong calculations. Meeting the requirements of Occam's razor, this Second Arrhenius Equation uses carbon dioxide as the only predictor. We can keep this formulation in mind as we explore the definitive anti-razor to this approach.

And that would be the modern system by which terabytes of information are acquired, assimilated, modeled, extrapolated, interpolated, mapped, and basically crunched to produce the weather forecast summarized with those little sun or cloud icons and predicted temperatures that make up modern weather forecasts.

I find it very interesting, and a little bit ironic, that this story takes us back once again to the Stockholm Högskola and the Stockholm Physics Society. We met Vilhelm Bjerknes in chapter 1 as a core member of the society. Like Arrhenius, he began as a lecturer at the Högskola in mathematical physics in 1893 and became a professor in 1895. He is credited with deriving the first "primitive" (meaning simplest) equations describing the physics of atmospheric dynamics in response to gradients in pressure, and so essentially defining many of the fundamental calculations that are carried out today in weather forecasting models. The stimulating environment of the society is, in turn, credited with allowing Bjerknes to step away from the observational methods generally used in meteorology at the time and to think in terms of fundamental relationships that could be described mathematically.

Those mathematical expressions are at the heart of what has become the most data-intensive and computationally complex area of environmental science—models of weather and climate. How did this happen?

Computer-driven models drive predictions of both your daily weather and our collective decadal changes in climate. But I sense in presentations some confusion about what models are and how they work. So let's spend a little time on this.

A model is a simplified representation of a real thing. Models can be physical representations. A traditional globe (figure 2.1) is a two-dimensional representation of the Earth spread over a sphere. Physical scale models of housing projects or redeveloped city streets are models of those imagined future realities.

One of my favorite physical models of the solar system is the Sagan Planet Walk in Ithaca, New York. Along a series of city streets, displays are to be found describing the size and nature of each planet. The physical location of each display tracks the relative distance of each planet from the sun. At the scale used for the walk, the sun and the first four planets (Mercury through Mars) are contained in one city block. Neptune is more than a kilometer away.

Small-scale physical models have been used in science to test, for example, the pattern of water flow through a valley (or watershed). My university houses a wave machine for testing modeled ocean turbulence on research equipment. There are lots of other examples.

Most models, however, are mathematical rather than physical. The GPS (Global Positioning System) or maps program on your phone is a mathematical representation of the road system in your search area. When you hit Directions, the model calculates different paths for you and picks the one that best fits your criteria—like "shortest distance" or "shortest travel time" or "avoid tolls." If your

Figure 2.1. A globe is a physical model of the Earth.

phone accesses real-time information on accidents, construction sites, and so forth, you are working with a system that is, in a small way, the same as the weather models in that it assimilates the most current information available.

Computational models combine equations representing what is being modeled in an effort to describe how that thing responds to change. So models are just a way of capturing how something of interest works, expressed in numbers and equations. Most often, the value in models is using them to predict how that something might behave in the future.

As an example, I do this at home with a spreadsheet of our household income and expenses to see where we have been and, I hope, where we are going.

Models can differ in terms of the time and space scales on which they operate. Our home financial model captures money flows monthly but tries to predict our financial future at a yearly time step (a time step is the frequency with which the calculations are made). The monthly model keeps track of income and expenses in lots of categories, but the annual model combines these into a single income stream and just a few expense streams (thus the scale or complexity or degree of detail of the model is different at the different time steps).

As with all models, predicting the future requires some assumptions. Basic assumptions for projecting a financial model into the future could include expected changes in income, date of retirement, return on investments, if any. Anticipated expenses could include basic costs for utilities, food and such, and longer-term expenses like rent, car loan, mortgage, and student loan repayments. Those would be called the driving variables or parameters; the outside forces that cause the model's predictions to change. The model itself is just bookkeeping. Funds move from column to column (box to box) as directed by the parameters.

And as with all models, those assumptions can be very wrong! That does not diminish the value of going through the process and making the best possible plans and decisions, based on the information you have. This model also places high value on trying to be as accurate as possible in estimating those driving variables. One thing you can do with a model like this is to project the future based on different assumptions. This is called testing "scenarios." For example, what if you retire at the age of sixty instead of seventy? What if you pay off that loan early, or take another one? Maybe you will move to a different city, changing both income and living expenses. Plug those into your spreadsheet and see what happens.

How does the Second Arrhenius Equation fit this description of a model? The driving variable globally is carbon dioxide concentration in the atmosphere. The predicted response or projection is global average temperature. The time scale depends on how often you measure the driving variable; annual changes would seem reasonable. I present the results of this set of calculations in chapter 11 using data that Arrhenius could only have dreamed of having!

The extent to which this equation captures changes in global temperature can be described as an Occam's razor explanation of climate change, but you would not want to use this equation to predict tomorrow's weather or to warn of potentially damaging weather events in the near future. Neither the time step nor the spatial scale is appropriate. The equation contains nothing of the deep understanding of the energetics and complex structure of the atmosphere that control the weather. Predicting tomorrow's weather requires a much more complex model. How complex?

The atmosphere is a very dynamic fluid that changes rapidly and varies in basic characteristics, like temperature and humidity, over very short distances. To capture and predict changes over hours to days, weather models need to represent the atmosphere in many, many small boxes and to do calculations over short periods of time. Smaller and faster are better, but how far this can go depends on how fast your computers are, and how accurate your measurements are of the current state of the atmosphere.

As an example, let's say we have divided the atmosphere into sixty-four thousand units or boxes (you will see why in a minute). If we picture the entire globe covered by these boxes (figure 2.2), you could divide the atmosphere into, say, ten layers from the surface to the top, and spread sixty-four hundred boxes per layer across the

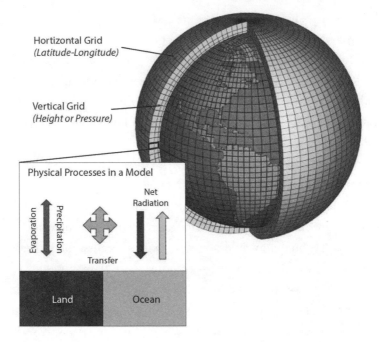

Figure 2.2. A conceptual diagram of the structure and scale of a weather prediction model. The surface is partitioned into a number of two-dimensional cells overlain by a column of boxes representing the vertical structure of the atmosphere. The size of surface cells and the number and depth of layers in the atmosphere vary among models, but the total number of boxes has increased dramatically over time as the power of computers has increased.

roughly 500 million km² of the Earth's surface. The average size of a box on the ground would be about 78,000 km², or a square about 280 km on a side (about the size of South Carolina or the Czech Republic), with nine more boxes of the same size stacked on top.

Why layers? Arrhenius did not have atmospheric layers in his model but used only surface temperature and predicted only a change from initial conditions. Weather is much more complex than a single temperature, and what happens in the upper atmosphere can have a huge impact on what we feel at the surface. A big part of that effect is caused by upward and downward movement of packets of air, so you need to describe the atmosphere in all three dimensions.

Given your sixty-four thousand boxes and the "primitive" equations of Bjerknes, it is then a simple matter to do the calculations for each box, determine how each box affects the adjacent boxes, and run this whole process forward at a speed that allows you to predict tomorrow's weather before it happens. Simple in concept anyway.

That sixty-four thousand number was not a random choice. In *The Weather Machine* Andrew Blum captures the concept known as Richardson's dream. Communications in 1913 between our friend Bjerknes and Napier Shaw, director of the British Meteorological Council, included an idea inspired by Lewis Richardson, an English mathematician whom Blum cites as having an "unorthodox intelligence."

Having applied that intelligence to developing mathematical solutions to a number of engineering problems, Richardson was intrigued by the problem of weather and forecasting, and, when made aware of Bjerknes's theory, he wanted to apply his mathematical solutions to those equations. The story is a good one, involving inventiveness, embezzlement (not by Richardson), and a stint in the medical corps during World War I, and Blum tells it well.

The upshot was that Richardson solved the Bjerknes equations and could make predictions, but his first effort to produce a one-day forecast required "the best part of 6 weeks." In *Global Warming: The Complete Briefing,* John Houghton says that Richardson's methods, published in book form in 1922, were correct, but that the computational load required to do a forecast for even one day was prohibitive.

Hence, at one point Richardson dreamed of a computational "palace" in which five hundred people would fill a mathematical conference hall ready to do the kinds of calculations Bjerknes described, based on accessing real-time weather information from a set of geographically distributed weather stations to predict the future state of the atmosphere—tomorrow's weather.

By the time Richardson published his book on this topic in 1922, the required number of human calculators had grown to sixty-four thousand. Needless to say, this palace was never realized.

So as early as 1922 the ability to predict tomorrow's weather was limited more by the accuracy and detail of measurements of the current state of the atmosphere, and of the computational power available to process that information, than by the basic understanding of the physics involved.

This interaction between accurate measurement and rapid computing has only been amplified over the past hundred years. Increased understanding of the physical processes affecting each box in the weather model has added some complexity to the calculations, but the basic processes remain the same (see figure 2.2). Conditions within each box include temperature, pressure, and humidity. These are impacted by sunlight and other energy sources (net radiation), which drive changed conditions within each cell and can result in the formation of clouds and precipitation. Differences in pressure between adjacent boxes affect the rate of air movement in three dimensions (transfer), mimicking weather fronts and the formation of storms. Concentrations of gases and particles in each box (like the greenhouse gas carbon dioxide) can affect radiation balance and temperature. Interactions between the lowest box and the surface (heat exchange, evaporation, friction) are included.

There are still unknowns in these models relating especially to the dynamics of water vapor and cloud formation, but the same two basic limitations still constrain the accuracy of weather forecasts: the accuracy and distribution of measurements of the current state of the atmosphere (now greatly augmented by satellite data) and the computing power required to grind out tomorrow's forecast. I look at the accuracy of weather forecasts, and how better information and more powerful computers have improved that accuracy, in chapter 4.

Blum clearly captures the intensity of the work, the absolute deadlines for performance, the complexity of the challenge, and the keen competition among different groups in terms of the accuracy of their forecasts. He also notes that in 2015 at least one commercial company changed from a human-driven forecasting process, where trained meteorologists interpreted the data provided to make those forecasts, to a completely machine-driven process, with humans left to interpret the output of the models to the public. Weather presenters now often see their role not as developing forecasts but as interpreting the computer forecasts to the public, including especially potential weather warnings.

What is the actual scale of these models? Because the accuracy of the initial representation of the state of the atmosphere is key to performance, the number of boxes has grown substantially over time, requiring a parallel increase in the amount of computer power required.

So how many boxes? Houghton describes current models as having a ground resolution of no more than 50 km and at least fifty levels up through the atmosphere. I'm getting my calculator. A square 50 km on a side would cover 2,500 km^2, an area about the size of Luxembourg and smaller than the state of Rhode Island. Divide that into about 500 million km^2 as the surface area of the Earth, and you get about two hundred thousand squares. Add fifty layers in the atmosphere, and you are now at ten million atmospheric boxes. The actual size of boxes varies with location and height in the atmosphere and changes with different models, but you get the picture.

Smaller boxes will change more rapidly, so the time step of the calculations is now about ten minutes; six times per modeled hour or 144 times per modeled day. With additional variables added to the basic primitive equations related to cloud formation and other factors, the number of calculations gets very hard to imagine, but with ten million boxes modeled 144 times a day, even a small number of equations means tens of billions of calculations per day.

Sixty-four thousand human computers does not begin to define the challenge.

Blum also captures very well the huge amount of data gathered from weather stations and now weather satellites as well, plus aircraft, sounding balloons, and maybe in the future smartphones, and again the task beggars the imagination.

And yet, several different organizations carry out this incredibly complex task every day, accessing all the available data, combining that data stream with the latest model, and crunching the global weather forecast. The European model Blum describes is updated every twelve hours, so double that unimaginable data acquisition load. The organization that runs this model employs about 350 people, and that is only those accessing the data and running the model. The number of people engaged in collecting the basic information needed, or keeping the automated systems that collect the basic information in working order, is greater still and includes professionals from almost every country in the world.

Even then, the forecast is generated at about a 50 km resolution. If you live near a coast or in an area with rugged terrain, weather can change a lot within 50 km. So the groups that produce those little icons and temperature predictions for your town (which will differ from the one they give for another town 50 km away) have to use some basic trends in climate data across or around that 50 km wide square to give you those specific little icons.

And yet, ask anyone you know about the accuracy of weather forecasts, and the general response will be that they aren't very good more than a couple of days out. For extreme events like hurricanes or tornadoes, frequent updates from radar stations and weather satellites are required to help us prepare.

Why? The short answer is: chaos. Not the general definition of social mayhem but the mathematical concept. We will leave that here for now and move back up to Occam's razor and the power of simpler explanations, returning to chaos in chapter 4. For now, keep in mind the incredible complexity of the modern weather prediction system, and still how the accuracy of the forecasts produced fades more than a few days into the future.

Which leads to this logical question: If we can't predict the weather more than a few days out, how can we predict climate change over the next hundred years?

The answer is this: Different forces control weather and climate at different time scales. Well, that is not very satisfying, is it? Sounds like a fudge. What does it mean?

Let's go a little deeper into the modeling process. Staying with our example of the household financial spreadsheet, I mentioned that the monthly sheet kept track of several categories of expenses, like automotive expenses, utilities, maintenance, credit card expenses, and so forth, but that the annual sheet grouped these into fewer categories. We could go deeper into the monthly expenses and subdivide utilities into electric, oil, cell phone, cable, ad infinitum. There are numbers for each of these, but to get a good sense of where the money is going, the category summaries are more useful and understandable.

We can do the same thing for climate models. What are the primitive equations? They represent fundamental laws of nature, right? Not really. Those fundamental laws are essentially statistical summaries of measurements at finer scales. You could try to describe them in terms of the basic energetics of molecular motion, but that would be pointless, since we know that the process is so well captured by the equations already in use.

Not to go too far into this, but suffice it to say that forces in nature are captured by statistical summaries of measurements of well-known physical (or biological or chemical) processes supported by a good theoretical basis. Essentially there is no fundamental level of explanation. Every process in any model (say, monthly utility expenses) is described using measurements and statistics applied at the next lower level of observation (oil, electric, cell phone), and if the model is successful, combining those processes allows prediction at the next higher level (the monthly budget).

In our discussion of climate models, the primitive equations are parameterized from observations representing more fundamental physical properties (the first level), which are then combined into the models we have been discussing (the second level) to make predictions about the future state of the atmosphere (the third level). Arrhenius had measurements of the effect of carbon dioxide on the absorption of longwave or heat radiation (level 1), which he then combined through extensive calculations to characterize the interaction of carbon dioxide and temperature (level 2), and then used this to predict changes in global temperature (level 3).

One point of this extreme pair of comparisons is that the level of complexity required to answer a question depends on the question. The Second Arrhenius Equation will not predict the weather for you, but we may not need the most complex weather models to predict or, perhaps more importantly, explain climate change, especially to general audiences. I come back to these topics in later chapters.

So where did this view of the modeling world come from?

I have been a forest ecosystem modeler for most of my career. There is an interesting dynamic in working as a modeler, being the person tasked with combining or synthesizing a wealth of data from very different field and lab studies into a single set of equations that captures the dynamics of a forest and predicts its future.

Forests are incredibly complex combinations of physics, biology, chemistry, and all the processes that build up from these, like tree growth and biodiversity, photosynthesis and decomposition. There are whole disciplines (and academic departments) built around each of these fields!

So how to decide what to put into your model—which equations to use and at what scale? This becomes the first question. There are many forest models out there. Some focus on carbon balances and so feature photosynthesis, decomposition, and the factors that control these processes. Others focus on the structure of forest canopies and how individual trees or species compete for light.

The only general answer to the question of model structure is: What do you want to predict? If carbon balance is your question, there are good reasons to use functional groups, like evergreen versus deciduous, rather than species. Or you can go even simpler and just use characteristics that affect photosynthesis, like the total weight of leaves in the canopy. On the other hand, if the diversity of species is what you want to predict, rates of photosynthesis might be secondary to other aspects of competition among species, like seed production and dispersal, tolerance of shade, and so forth.

Basically, any one model that tried to include all of the processes relating physiology and biodiversity at the same time would become ponderously large—unwieldy perhaps. I venture to say that it would become impossible to measure all the parameters for any one forest required to run such a model. So with a well-defined problem and scope in mind, the modeler can be put in the position of saying to colleagues, "Well, that is a really interesting piece of science, but it does not relate to the goals of this model, so we won't include it." That is not a popular stance!

This kind of critical thinking was sharpened while I was part of a research team that was trying to develop a satellite that would provide measurements of forests to be used in global models. What would you measure, and how?

My participation in this collaboration began when a NASA scientist became aware of the work my group was doing in measuring the chemistry of leaves using the reflectance of light instead of by traditional wet chemistry involving beakers, acids, and Bunsen burners. We were also working with forest models that could use that chemical information about leaves to predict forest dynamics at the landscape scale.

My NASA colleague-to-be asked this question: If you could design a satellite to provide information to drive your forest models, what would it be? My answer, offered in jest, was to launch a version of our laboratory instrument into space, so we could measure the chemistry of intact forest canopies and predict the rate of photosynthesis. It turned out that scientists at the Jet Propulsion Laboratory at Cal Tech in Pasadena, California, were working on exactly that kind of instrument, though it was intended for geological research, including prospecting for gold!

This connection led to several years of very rewarding interactions with geologists, chemists, engineers, and program managers at NASA and JPL on instrument design and performance.

To set the specifications for this instrument, the engineers (and geologists as well) wanted to know the fundamental variables needed to predict the carbon balance, or growth, of a forest. Quite a bit of the conversation focused on this question: Do species matter? From space, forests look like a green surface carrying out photosynthesis, and it's hard to tell an oak from a maple. One conclusion from this work, which formed the basis for a model we built to use the satellite data, was—no, not really. Major functional differences did matter: deciduous versus evergreen, total chlorophyll content (chlorophyll being the pigment or molecule in leaves that makes them appear green and captures light energy for photosynthesis). It turned out we could map the rate of photosynthesis based on a single measurement of the reflected light from entire forest canopies.

Question: What is the rate of photosynthesis in that forest down there? Measurement: reflected light related to the amount of chlorophyll in the entire canopy. Model: rate of carbon gain determined by that measurement (and climate variables). William of Occam might approve.

My experience with forest models leads to these questions about climate change: What is the right resolution for the model, and what are the most important measurements and parameters? If you like, this is asking the Occam's razor question: What entities are necessary and sufficient, and what can be excluded? We will leave that question hanging out there for now.

One more set of ideas about modeling as related to Occam's razor.

Some years back, I wrote an editorial for a professional journal describing what I saw basically as loose thinking among ecosystem modelers. How can I capture this for you?

Let's go back to your home financial model, and let's say you want to predict how much money you will have in the bank at the end of the year. You have all these numbers to enter about cash flows due to income and expenses, and you run your model but you don't get the right answer (as determined by your actual bank account or level of debt at the end of the year), or you don't get the answer you want!

Well, you can go back and change the numbers! Now we get into fuzzy territory. You could double your projected income. You could cut your rent or mortgage payment in half. You can do lots of things in a model that you can't do in reality. In general, in a model with lots of variables and only a single prediction (money in the bank), there are lots of different combinations of numbers that could get you to the right prediction.

The process of changing inputs to achieve a certain outcome is called calibration, but I prefer the term "tuning."

My editorial groused that, at least in ecosystem science, only modelers were allowed the luxury of changing the values of lots of parameters to get the answers they wanted. Field scientists were held to a higher standard: you needed to have many measurements of a process to draw significant conclusions about even that one process. The paper suggested that this was a bad situation, that the predictions from such models were meaningless unless the parameters were directly tied to measurements. In essence, this is another argument for Occam's razor, which could be expanded to suggest that we not add more "things" to the model if they can't be measured or defined.

This is a lot of words on a pretty theoretical topic. This is all included here because models are front and center in most presentations about both weather prediction and climate change, so some familiarity with the process should be helpful.

A takeaway message here is that models at very different scales or levels of complexity can be useful in both predicting the future and explaining climate change, for example, to general audiences.

I'll conclude this chapter with a favorite example of the value and power of simpler explanations.

In the original version of J. E. Lovelock's *Gaia* (1979), he offers a simple planetary-scale definition of life: the use of the sun's energy to produce and maintain chemical disequilibrium in the atmosphere. Sounds theoretical, but the idea had important implications.

NASA was in the throes of developing a program to determine whether or not life existed on Mars. Much effort was going into developing landers that could sample Martian soil, do basic biological and chemical tests, and look for compounds that would be produced by living organisms, at least as we know them on our home planet.

Lovelock said that, based on the fact that the Martian atmosphere was at chemical equilibrium (unlike our own atmosphere dominated by nitrogen and rich in oxygen, the product of photosynthesis), life did not exist there; a simple and elegant solution. Lovelock was not invited to continue to work on the project.

To bring this to a close, I am stating the extreme case for examining simpler answers to complex questions. This is not intended to diminish the importance of those high-end, complex models. They represent cutting-edge science, and we need them to summarize and integrate what we know about weather and climate, and to suggest where more research is needed.

I am making the case for simpler models that can capture the most important dynamics of long-term climate change in ways that are clearer to general audiences. Perhaps this can help build confidence in the conclusions drawn from the science behind both complex and simple models.

So the first room in our guided tour of weather and climate has presented two very different ways of approaching the science: the Second Arrhenius Equation and the modern, complex weather model. Occam's razor might reach across the centuries to ask how far we can go in explaining climate change with fewer things rather than with more. But before we can address that question, we need to understand the basic processes driving weather and climate.

From Weather to Climate

DYNAMICS IN TIME AND SPACE

3
The Basics of Weather and Climate—It's Simple!

There are so many different ways now to experience weather! You can walk outside and see what is going on, but you might be more likely to check your phone first. There is so much information available in an instant that a simple question such as "Is it raining?" gets not just a yes or no but a quick look at a radar app that leads to answers such as "not yet" or "it just stopped but not for long." Those comments capture the basics of weather prediction. What can you see that has happened in the past hour, and what do you predict for the next hour?

We have looked at two very different examples of ways to predict weather or climate, neither of which has conveyed much about the basic forces or processes that actually make weather happen. So now let's go deeper than Arrhenius but also back away from the ten-minute, multibillion calculation complexity of weather models. Let's focus on the forces that cause the changes you see on that phone app, and explain them with some basic facts, most of which you already know.

Fact 1: The shape of the Earth and the tilt of its axis (figure 3.1), along with the annual orbit around the sun, result in very different amounts of energy received per unit area of land or water at different times and latitudes. At the equinox, when the Earth's axis is parallel to the sun, the curvature of the Earth causes the flat plane of solar energy to be spread over a larger area at the poles than at the equator, and total energy received per unit of land or ocean area drops dramatically

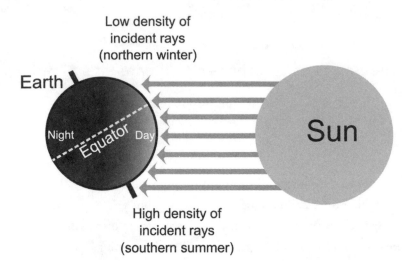

Figure 3.1. Solar energy is unevenly distributed across the surface of the Earth due to the curvature of the surface and changes in orientation to the sun over the course of an orbital year.

as you move away from the equator. As the Earth goes around the sun during the year, each pole points toward and then away from the sun, creating the annual cycle of seasons and accentuating the differences between the equator and the poles.

Fact 2: The weather/climate machine is a single giant mechanism attempting to redress this unequal distribution of energy. This happens most rapidly in the atmosphere, which responds quickly to these received inequalities. Weather, the topic for this and the next chapter, happens in the atmosphere. Climate, or changes in weather over years to centuries, is also strongly influenced by energy redistribution and storage in the oceans and in ice, parts of the Earth system that change much more slowly than the atmosphere.

These two facts cause temperatures to be higher on average, and much less variable across the year, in the tropics than at the poles.

Fact 3: Hot air rises, and cold air falls. This is because warmer air is less dense and so will tend to rise above, or through, colder air. If you want to demonstrate this phenomenon, turn on a burner on your stove or hot plate and feel the warm air rise past your outstretched hand.

Fact 4: Nature abhors a vacuum, so whenever air moves away from one location, as when warm air rises from a surface, that will draw air from an adjacent location to replace it—creating a circular motion or a cycle of air movement. I keep going back to Arrhenius and the Stockholm Physics Society, but this was one of the advanced and insightful realizations that supported much of their innovative discussions—that cycles of all kinds are an inherent property of all processes in nature.

Combining facts 1 through 4, warm air rises over the equator and has to be replaced by surface air moving in from more northern or southern locales. Then where does that air come from? At some point, that risen tropical air has to come back down and replace the surface air moving toward the equator.

Do you sense a cycle here? In the simplest terms, hot tropical air should rise to the upper atmosphere, move toward the poles (cooling as it does), and at some point should return to the surface. And that is (almost) exactly what happens. In reality, that simple, single cycle is broken up into three adjacent cycles, or cells (figure 3.2). Warm tropical air tends to descend about one-third of the way to the poles, or about 30 degrees north or south of the equator. This circulation pattern gives rise to what have been named the Hadley cells.

Fact 5: Where adjacent cycles meet, they will move air in the same direction. So how does this 30 degree cycle relate to other redistributions of unequal solar energy? There is another set of roughly 30 degree cycles (called Ferrel cells or mid-latitude cells) to the north or south of the Hadley cells. Air at the equatorial end of the Ferrel cells descends in concert with the polar end of the Hadley cells (figure 3.2).

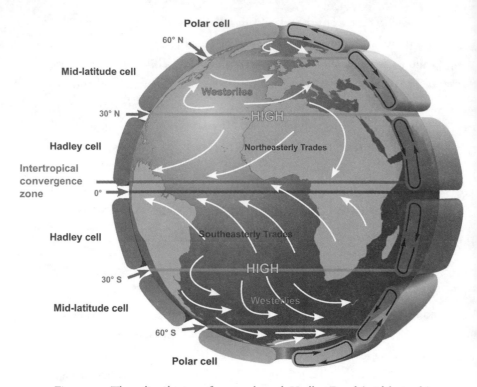

Polar cell
60° N
Mid-latitude cell
Westerlies
30° N
HIGH
Hadley cell
Northeasterly Trades
Intertropical convergence zone
0°
Hadley cell
Southeasterly Trades
30° S
HIGH
Mid-latitude cell
Westerlies
60° S
Polar cell

Figure 3.2. The redistribution of energy through Hadley, Ferrel (mid-latitude), and Polar cells, with resulting zonal wind patterns. Deflection from direct north-south wind flows results from the Coriolis effect.

And where does the descending air in the Ferrel cells originate? At roughly 60 degrees north and south. Rather than my going on here, you can see how this plays out in figure 3.2. In general, we can describe three interacting cycles: Hadley cells, Ferrel cells, and Polar cells.

This simple description of energy redistribution has major implications for the distribution of ecosystems, vegetation types, and agriculture, and for human history as well. To explain why, we will have to bring humidity into the picture now.

Fact 6: Air with a lot of moisture in it (humid air, as on a hot, sticky summer day) is even less dense than dry air and will rise faster. Compare your hand over the dry burner with a hand over a pot of boiling water! Compare the warmth you feel over the boiling water with the warmth over the empty burner, and you will also get a sense for how much more heat can be transferred up by humid air than by dry air. I come back to this when we consider local and daily weather patterns later.

Fact 7: The amount of water vapor (or humidity) that air can hold is related to temperature. This is a little harder to picture, but this fact controls patterns of evaporation and condensation that relate to everyday occurrences.

When you hold a hand over boiling water, you will feel the transfer of heat, but your hand will also become moist. The air leaving the surface of your boiling pot is saturated with water vapor—100 percent relative humidity. As the air reaches your hand, it transfers some of that heat to your hand and becomes cooler. Since cooler air can hold less water vapor, the vapor condenses on your hand.

Evaporation and condensation can be seen all around. A cold drink, in a cold glass, on a hot, muggy summer day becomes covered with water. The air next to the glass is cooled by the glass and so can hold less moisture, which is why the moisture condenses on the glass. The dew that forms on your car or lawn after a clear summer night results from a similar process. The surface cools faster than the air, so the air near the surface cools, and some of the moisture in the air condenses on the surface.

The evaporation and condensation of water vapor is a crucial part of all weather and climate patterns. The energy absorbed or released by these processes is very large and can play a major role in things like thunderstorm formation. Again, more on that a little later.

Warm tropical air tends to be very humid as well, and as it rises, it cools. As it cools, it can hold less moisture, so clouds tend to form and rain tends to fall. The moisture is squeezed out of the air, if you like, and returns to the ground or ocean around the equator as rain (think of a tropical rain forest).

At the other end of the cycle, at 30 degrees north or south, the air warms as it descends, so it can hold more moisture, but without a source of water vapor, the air just becomes drier (less humid). Interesting maybe, but how does that relate to climate and ecosystems? There is a tendency for deserts to form at 30 degrees north and south of the equator, as a result of this descending warm, dry air. The Sahara, Namib, and Kalahari Deserts in Africa and the Mojave Desert in the southwestern United States occupy these zones.

A sharp observer might note that this means that the southeastern United States should be desert as well—and it is not—an example of how these general trends can be offset by local circumstances. The proximity of the American South to the Gulf of Mexico and the tendency for storms to draw on this source of moisture create a local increase in rainfall that offsets the global pattern. Seasonal monsoons drawing moisture off the Indian Ocean play the same role in Asia.

Rising humid air around 60 degrees north again cools as it rises and tends to lead to greater rainfall, more humid conditions, and more forests and agriculture, especially across Europe and North America.

So with a few basic facts related to solar energy received and the tendency of humid and dry air to rise and fall, we can begin to explain the distribution of major climate patterns and even ecosystem and vegetation types, with the caveats described above.

What does this say about the average or normal pattern of winds at the surface of the planet? Looking at the cells on the exterior of figure 3.2, it looks as though winds should go north to south between 30 degrees north and the equator, and south to north between 30 degrees and 60 degrees north, but the actual wind patterns depicted by the arrows across the surface are more complex.

Do you enjoy a good sailing adventure? My favorites are the Aubrey/Maturin tales penned by Patrick O'Brian. In either fiction or history, the age of sail required deep knowledge of global wind patterns. The route from Europe to North America might involve first heading south to pick up the tropical trade winds blowing east to west. From North America to Europe, captains might try a northern route to catch the "westerlies," which generally blow west to east, even at the risk of colder weather and possibly fierce storms.

If average wind directions should be primarily north and south, as suggested by those nice circles on the perimeter of figure 3.2, then why are the actual flows at the surface more east and west?

Fact 8: The rotation of the Earth deflects these north/south winds through what is known as the Coriolis effect. While the winds are blowing "vertically" (north and south), the Earth is spinning under those winds. To those of us tied to the surface, the winds then appear to be moving more "horizontally" (tending east and west).

Perhaps you have seen a video that captures this effect. If not, there are some really good ones on the Web. In the video, two people are on opposite sides of a playground merry-go-round and begin to spin. There are two cameras in operation, one on the merry-go-round and one on the ground outside it. One person has a ball and throws it toward the other.

The camera on the ground records the ball flying in a straight line, but the second person can't catch it, because the spinning merry-go-round has moved them away from the spot where the thrower aimed the ball. The camera that is on the merry-go-round records how this looks to the person throwing the ball. To that person, it looks as though the ball has curved away from the receiver. In terms of average wind direction on the surface of the Earth, we are the person throwing the ball!

In the simplest sense, this rotation of the Earth "under" the prevailing direction of wind movement within one of the cells in figure

3.2 deflects average wind direction to the east in the Ferrel cells and to the west in the Hadley cells—yielding the trade winds and westerlies (I have always found it confusing that the westerlies blow *from* the west, not *toward* the west).

It is the Coriolis effect that also causes major storms to rotate in different directions in the northern and southern hemispheres (you can see this on any global image of current cloud formations). Low pressure centers associated with storms rotate counterclockwise in the northern hemisphere and clockwise in the southern.

Now for one more important global pattern in the weather/climate system. If you have had the chance to take a long-distance, round-trip flight within the northern Ferrell cell (say, Chicago to Paris or Los Angeles to New York), you will have noticed that the trip east takes a lot less time than the trip west. If you get to go more than once, you might notice that two trips of identical distance might take very different amounts of time. Why?

Fact 9: Where the Polar, Ferrel, and Hadley cells converge, dynamics at the boundary create fast-moving streams of air known as jet streams. These aptly named high-elevation, high-speed rivers of air are the aviators' version of the old sailors' trade winds. Pilots, within the limitations set by the rules of flying and approved routes, might try to find or avoid the jet stream, depending on which way they are traveling. So what are jet streams, and how do they affect air travel?

Well, we are back to the Hadley, Ferrel, and Polar cells depicted in figure 3.2. That figure might seem to describe a smooth and easy blending of upward and downward airflows, but a closer look gives a different impression (figure 3.3). Tropical air is warmer than temperate zone air, which is warmer than polar air. The extra energy causes warmer air masses to lift higher into the atmosphere. Differences in height and differences in temperature between either the Hadley and Ferrel or the Ferrel and Polar cells lead to turbulence when these two

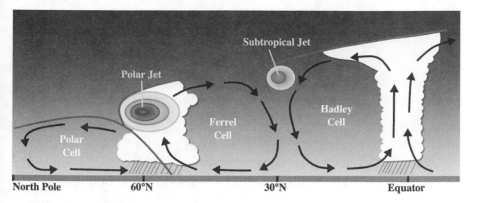

Figure 3.3. Turbulence at the intersection of Hadley, Ferrel, and Polar cells, in combination with the Coriolis effect, leads to the creation of the polar and subtropical jet streams.

airflows converge, producing a rapid spinning motion. Combined with the Coriolis effect, these narrow, high-elevation "tubes" of air are sent speeding west to east as they spin.

In general, the polar jet, between the Ferrel and Polar cells, is stronger than the subtropical jet, between the Hadley and Ferrel cells, and can exceed 150 km per hour. It's not hard to imagine that such strong, narrow wind fields will be turbulent at the edges, and so that fast flight to Paris might not be the smoothest.

That seems to be a nice static pattern in figure 3.3, but the jet streams are anything but static. Several websites are available that depict either recent history or future predictions of the location and speed of the jet streams. What you see is an incredibly dynamic interface among the three idealized air masses projected in figures 3.2 and 3.3. The animations will show that jet streams can speed up or slow down, shift far north or south, and sometimes even combine across our idealized "cells" as jets merge and separate. Predicting the high-elevation wind patterns at these interfaces is important for weather

prediction at the surface, as the energy and turbulence at those interfaces can drive important weather events.

In particular, how "wavy" these streams of air are plays a major
role in the severity of storms we experience, especially at mid-
latitudes. Figure 3.4 compares two different patterns in the polar jet
stream. Most simply put, when the polar jet depicted here flows
straight across, there is less interaction between the warm and cold air
masses on either side of the jet, and fewer storms. When that jet is
wavy, those interactions are stronger, and major storms can occur.

There can be several waves in the polar jet at the same time, as
you can see in figure 3.4. These are called Rossby waves. They generally move west to east, hence the general movement of storms, again
at mid-latitudes, is also from west to east. How this relates to outbreaks of very cold winter weather in the northern hemisphere is a
topic for chapter 5 when we visit the polar vortex.

We talk about jet streams casually now as if this is something
we've known about all along, but consider how difficult it would have
been to experience or identify this basic atmospheric phenomenon
before the advent of high-elevation aviation. It is intriguing to me
that such an important part of our understanding of the global
weather system remained unknown until the late 1930s.

Step back for a minute here and realize that all of the complexity
in these descriptions of global weather and climate patterns result
from the nine facts listed, many of which you already knew.

They also drive important regional patterns that can be modified
by local conditions. If the Gulf of Mexico preserves the American
Southeast from being a desert, are there other geographies that determine weather and climate? There are, and we can see that they determine vegetation and agriculture as well.

Here is one example. If you have the chance to drive around the
Olympic Peninsula in western Washington State on U.S. 101 as far as

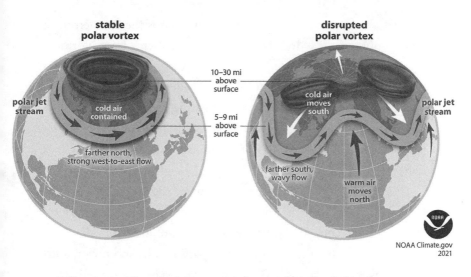

Figure 3.4. The polar jet stream can be straight or "wavy," with major implications for regional weather.

Olympia, then take Interstate 5 north up to I-90, and then go east toward Spokane, you will pass through an astounding array of climates and vegetation types.

The Olympic Peninsula supports one of the most luxuriant stands of temperate rainforests in the world. Spruce, Douglas fir, and hemlock trees soar hundreds of feet in the air, while the forest floor is covered with mosses, ferns, and decaying logs. The western edge of the peninsula faces into perennial onshore breezes off the Pacific Ocean, which provide abundant rainfall and tend to keep temperatures moderate. Images on the U.S. Park Service website for the Hoh Rain Forest give a glimpse of the immense accumulations of living trees, dead wood, mosses, and ferns supported by this wet, temperate climate.

Driving south along the coast on U.S. Route 101, and then east toward Olympia, Mount Olympus will be to your left, although you may not see it for the clouds. Mount Olympus is the northernmost outpost of the Coast Mountains that hugs the Pacific from Northern California to your current location. Continuing on 101 east, you pass through the Coast Range and into the valley hosting the cities of Tacoma and Seattle. Changing in Olympia to I-5 north and then I-90 east brings you within sight of Mount Rainier, and up and over the Cascade Range running north to south inland from the Coast Range.

Forests in the Cascades contain many of the same species as the Hoh Rain Forest, but the climate is drier and the growth less vigorous. When you come down off the eastern side of the Cascades, things look entirely different. Welcome to the grasslands, and now mainly farmlands, of the Palouse. The climate has become distinctly drier, and the forests have largely disappeared. Why?

Imagine a parcel of warm, humid air driven off the Pacific Ocean toward the Olympic Peninsula by the prevailing westerly winds. Reaching Mount Olympus (elevation 7,980 feet or 2,425 meters) and surrounding mountains, the air is forced up and over the ridge. As the air rises, it cools, and as cooler air can hold less moisture, condensation leads to cloud formation and rainfall—lots of it! Average annual precipitation at the peak of Mount Olympus can reach more than 180 inches (450 cm) per year.

Passing over the Olympics, the air, now somewhat drier, descends and warms toward Seattle, where average annual rainfall might reach 40–60 inches (100–150 cm). Continuing to the east, our air parcel now encounters the Cascade Range (topped by Mount Rainier at 14,411 feet or 4,380 meters) and is forced up again. Once more the air cools, water vapor condenses, clouds form, and rain (or snow) falls. Total precipitation in the range averages 80–100 inches (200–250 cm) per year and can reach 100–140 inches (250–350 cm) at the highest elevations.

When our air parcel crests the Cascades and descends to the Palouse region, most of the moisture has been "squeezed out," and as the air warms on descent, relative humidity drops. The driest grassland regions of the Palouse might receive only 10 to 20 inches (25 to 50 cm) of rain per year. Basically all the moisture coming off the Pacific has been lost by its passage over these two impressive mountain ranges.

We have just described the impact of the rain shadow effect, the name given to this process of draining moisture out of the air as it passes up and over mountains (the technical term is orographic lift).

For another set of local and regional examples, let's ask this question: What do the South Asian monsoon and the relief you feel at the beach on a hot summer day as cool air comes off the ocean have in common? The answer: They are both the result of geography and differential heating—a good example of the same basic process happening at vastly different scales.

Let's consider the sea breeze first. Have you noticed that refreshing sea breezes tend to be strongest in the late afternoon? If you are thinking this has to do with differential heating by the sun, you are correct.

Water has a huge capacity to absorb energy without changing temperature, which is just a long way of saying that the water temperature at your beach does not change much during one day. If you try to swim in the ocean off Maine or Washington State, you know it doesn't change much over the course of a whole summer either, but that involves ocean currents, and we are not going there yet.

In contrast to the ocean water, the surface of the land heats quickly under full sunlight. You experience this as you hop over the hot sand on your way to the water. Now you can predict what is going to happen. The heated air rises, pulling the cooler air off the water, to your great relief. The part of the cycle you can't feel is the movement of the risen warm air out over the water, where it cools and eventually descends back to sea level, completing the required cycle.

How does this small-scale circulation relate to the continental-scale, seasonal monsoons in India, a phenomenon crucial to crop growth and the production of food for a population of more than 1.3 billion people?

India south of the Himalayan Mountains is relatively flat and at low elevation, especially relative to the world's highest mountains to the north. Due in part to a rain shadow effect, the Tibetan Plateau north of the highest Himalayas is a vast, dry semidesert area, with annual precipitation ranging from near zero to as much as 38 cm per year.

One aspect of deserts is that, because they are so dry and the air over them is so dry, they experience huge ranges in temperature over the course of a day or a year. In summer, heating of the plateau creates an effect you can now predict. Hot air rises, drawing cooler and moister air off the Indian Ocean. As this responds to uplift caused by topographic features in the Indian lowlands, and then finally at the foot of the Himalayas, the crucial rainfall follows.

Although this characterization describes an average condition, the timing and amount of rainfall occurring during the monsoon season vary widely, with huge consequences. The monsoon can also be affected by interactions with other large-scale climate patterns, which we will encounter in chapter 5.

This chapter started with us checking our phones for the local radar. Let's wind it up by going back to local events. We can finish with a flourish by describing some dramatic scenarios, and how they relate to the jet stream, but let's start with just some typical daily patterns.

Commercial weather sites might show you a panel of predictions reaching up to ten days into the future. We may tend to look primar-

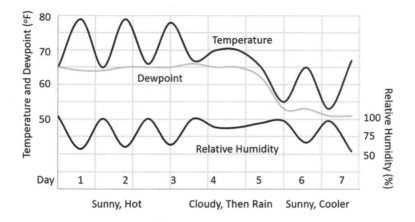

Figure 3.5. A hypothetical seven-day local weather pattern showing the interactions among temperature, dewpoint, and relative humidity.

ily at the sun/cloud/rain icons and the maximum and minimum temperatures in those forecasts, but some of the more complete graphics demonstrate a number of the interactions described in the regional and daily patterns presented so far.

Figure 3.5 presents a hypothetical example of the changes in temperature, dewpoint, humidity, and weather over a seven-day period. So it is summer, and you start your vacation with three sunny and warm days. The top line shows the daily change in temperature for those first three days. Temperature rises and falls as expected, and the difference between the high and low is large because cloud cover is low and a lot of solar warming is happening.

The bottom line is relative humidity. We hear this expressed usually as a percentage (as in relative humidity is hovering at an uncomfortable 85 percent today). But a percentage of what?

Here we need to introduce the concept of dewpoint, or the temperature at which the air will be saturated with moisture (recall fact

7—that the amount of water vapor air can hold increases dramatically with temperature). Dewpoint temperature is determined by the actual amount of water vapor in the air, or the "absolute humidity." Relative humidity is expressed as the amount of water vapor in the air as a percentage of the amount that could be held at the current temperature.

So in figure 3.5, over the first three days, the dewpoint, or the actual amount of moisture in the air, changes little. As temperatures rise in the morning, relative humidity drops, as that warmer air could hold more moisture. At the end of the day, the absolute humidity and the dewpoint haven't changed, but it may feel increasingly humid to you as the temperature drops and relative humidity increases.

Then what does cause a change in dewpoint?

After those first three warm summer days, the weather (in figure 3.5) changes. Cloud cover increases on day 4, leading to rain on day 5. Temperature does not increase as much as it did during the first three days, as there is less sunshine, so relative humidity stays higher as well. Rain often means a change in the air mass in place, and in this case, that warm air has been replaced with cooler and "drier" air. Drier air in this case means less absolute humidity, less actual water vapor in the atmosphere, and a lower dewpoint temperature.

In this figure, actual temperature is always above the dewpoint. Is that true in general? The answer to that question relates to facts 10 and 11, which round out our complete set.

Fact 10: It takes a lot of energy to evaporate water, and a lot of energy is released when water vapor condenses. This is one reason perspiration or a splash of cool water on your face cools you off on a hot day. The evaporation of that water is drawing heat energy off your skin. On the other hand, that condensation on your cold glass of water is warming the glass and its contents.

Fact 11: Because of the energy released by condensation, nighttime temperatures can never fall below the dewpoint temperature. At that temperature, condensation will occur, releasing energy as water vapor turns to liquid. Rather than getting lower temperatures, you get the formation of dew, and the energy released "warms" the air, keeping it at the dewpoint temperature. This is why hot, sticky summer nights stay hot and sticky, even when the sun is down and the sky is clear. We could rightly change the frequent statement "It's not the heat, it's the humidity," to "It's not the heat, it's the dewpoint!"

The other side of Fact 11 is that if you live in a dry climate, where humidity is always low (think deserts and dry grasslands), you can experience very large swings in temperature. Clear skies will drive temperatures up during the day, and a very low dewpoint will allow temperatures to drop rapidly right up until dawn. Temperatures may never reach the dewpoint.

Shifting away from the humdrum of calm and peaceful weather days, let's look at winter snowstorms and summer thunderstorms with tornadoes—two of the most dramatic and violent kinds of weather events we experience. Both can relate to how our simple weather rules interact with the jet stream.

Let's start with simple summer thunderstorms that can form without any help from the jet stream.

You have a long day off, and it is going to be a triple-H day (hot, humid, and hazy), so you set up on the porch with a glass of lemonade and a tube of sunblock. The sky is clear, if hazy, and the sun is hot. The air is still, the atmosphere is calm, and there is no wind. As a weather expert (and because you have to avoid stepping on the deck in bare feet), you know that the strong sunshine is warming the surface. As the day goes on, the air near the surface warms as well, and you start thinking about convection. (What is convection? Just a technical term for the fact that warm air at the surface wants to rise up through cooler air above—fact 3.)

Around noon you start to notice clouds appearing and you are grateful for a little shade. As the cloud cover increases, you begin to feel breezes. As the warm surface air tries to rise, it does so in fits and starts, with local uplift and replacement near the surface. A weather forecaster at this point might say that the winds are light and variable.

By midafternoon, the cloud cover is nearly complete, and the breezes are stronger. This all feels like relief, but the temperature is still high, as is the humidity.

Suddenly the clouds on the horizon start to look ominously dark. Around 5 P.M. you start to hear thunder and you think about heading inside. Just before sundown, the winds start to howl (that is the technical term), and thunder and lightning are all around. The rain pours down and lowers the temperature by 10–15 degrees. Relief at last, and you have a comfortable night's sleep.

So, what happened here? Strong sunlight warmed the surface, and the warm humid air wanted to rise through the cap of cooler air above. This is not a smooth process. Convection is by nature a turbulent process. Pockets of heated air "bubble up" through the cooler air, causing light and variable winds at the surface; variable in terms of direction. This is just like the bubbles rising through the water from your pot on the stove as it starts to boil.

As the hot, humid air rises, it cools. Water vapor in the humid air condenses, forming clouds and then rain. Remember fact 10— condensation releases energy, and this extra energy can give the cloud formation an extra boost, bringing the cloud top higher into the atmosphere. If the uplift is extreme and the moisture freezes, it will come back to Earth as hail—take cover! When water freezes, more energy is released, giving the growing storm another boost! The downward flow of rain or ice to the surface, along with increased turbulence and the downward movement of cooler, higher air, creates cooler surface temperatures, bringing you relief and shutting down the convective uplift.

You are all set until the same process happens tomorrow. If this sequence is not complete, and there is not enough convection or moisture to form thunderstorms, you are in for a hot, sticky night.

That simple summer day can become more complicated, and more dangerous, if the jet stream gets involved.

So, let's change the scenario. There you are on the deck in the heat and humidity (and sunblock). Instead of a few puffy clouds forming at midday, you see on the horizon a full bank of very dark, very tall clouds. It is coming toward you. Suddenly your phone, radio, tablet, TV, or whatever starts to beep madly, and you receive a severe storm warning, or perhaps a tornado warming. You head inside right away and get to the basement, or whatever safe place you might have, and wait for the storm to break. And it does, with dangerous winds and torrential rain or maybe hail. If you still have cell phone service (or a TV in the basement and the electricity is still on) you watch the radar image, as a bright red storm front graphic passes overhead. When it is gone, you come out and assess the damage.

What was different? Instead of convection being a local process as in the first scenario, this time a weather front being driven by the progress of a Rossby wave causes massive uplifting of that warm humid air and a concentrated, violent mixing of warm and cold. Clouds can reach up to the height of the jet stream, and if building clouds reach that high, the jet stream can give them a big spin, which can translate all the way back down to the surface in the form of deadly funnel-shaped clouds—tornadoes!

On the plus side, that moving front probably brings with it much cooler and drier air from the Polar cell behind it, so tomorrow might be altogether more comfortable than today.

Then what about those big winter storms? In the eastern United States, the most intense winter snowstorms generally track the jet

Figure 3.6. A visible image of the intense nor'easter of January 2018.

stream and occur when the polar jet is "wavy" due to a Rossby wave, as seen in figure 3.4.

To get a real northeaster (or nor'easter as New Englanders like to call them) a low-pressure system follows the jet stream from southwest to northeast, and just off the coast. As the storm swirls counterclockwise, relatively warm and very moist air is drawn off the Atlantic Ocean and feeds the storm with energy and moisture. An intense nor'easter can almost look like a mini-hurricane (figure 3.6).

Tornadoes, monsoons, snowstorms, rain shadows, rainforests, deserts—all the result of a few simple rules matched with geography and sunshine. Fairly simple in general, although the details of those major storms are wildly complex and bear continuous watching. It is impossible to overstate the value for our own safety of the accurate imaging and short-term prediction of the progress of major storms available to us from the complex weather monitoring and prediction process described in chapter 2.

Beyond immediate storm tracking and today's forecast, predicting the weather next week is another matter entirely. While the same basic rules still apply, the accuracy of detailed weather forecasts declines quickly as we look even a few days into the future. What can explain this apparent contradiction?

4

If Weather and Climate Are Simple, Why Is Prediction So Hard?

We are awash in numbers. They define who we are officially and how we conduct the basic stuff of life. I have more than three pages of passwords just for all the professional, financial, and commercial websites I use, and then there are your Social Security number, phone number(s), even library cards, for heaven's sake.

A lot of the numbers we see every day describe performance. The outcomes of sports events are reported in voluminous tables describing just how well your team, and all of the players, actually did last week. Analytics have come to dominate reporting and strategy in baseball. In the business arena you have daily reports on stock prices, sales volume, profit, all set against predictions. Did this company beat expectations today, or this quarter, or this year? Performances of well-known financial advisers and pundits are published regularly. In entertainment and social media, numbers range from viewer opinions (think Rotten Tomatoes) to revenues, and now to subscriptions, downloads, hits, and likes.

Weather forecasting is also awash in numbers, but numerical analyses of performance are not part of the daily conversation. Forecasters almost always look forward rather than looking back to talk about how accurate their previous forecasts were. Predictions seem to change several times a day, and in general we don't seem to expect accuracy more than a couple of days into the future; sometimes less. At the same time, the accuracy of severe weather warnings covering the next couple of hours, received from those same forecasters, are absolutely critical, and save lives every year.

If chapter 2 shows that the basic equations used for weather forecasting have been known for some time, and chapter 3 illustrates that local and regional trends, like rain shadows and sea breezes, are not hard to understand, then why are forecasts for more than a few days out so hard to make? And what does "a few days" mean? For that matter, what do we mean by "accuracy"?

Before diving in, let me summarize three major conclusions to be presented here.

- The accuracy of weather forecasts has been increasing steadily for several decades.
- Forecast accuracy declines rapidly as we look further into the future. For temperature, predictions made more than about ten days beforehand tend to be no better than using the long-term average for that day. Precipitation is even harder to predict.
- This will always be true, and the reason for this is chaos.

Are there numbers to support points 1 and 2? It takes some digging, but there are, and the results will lead into our discussion of point 3.

The U.S. Weather Service has extensive data on the accuracy of its forecasts under the title "verification" on its website. Figure 4.1 presents nearly forty years of data on mean absolute error for predictions of maximum temperature three to seven days into the future. Mean absolute error is the difference between predicted and measured recorded as a positive number, whether too high or too low. Days in which the prediction was 4 degrees too high or 4 degrees too low would both give the same absolute error: 4.

Annual WPC Mean Absolute Errors
Maximum Temperatures

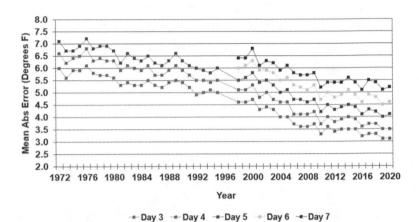

Figure 4.1. An indication of the improvement in prediction of daily high temperature over time by the U.S. Weather Service. The lines, from top down, are for predictions seven, six, five, four, and three days into the future. Mean absolute error is the average difference between predicted and observed temperatures, always expressed as a positive number. A prediction that is either too high or too low by 6 degrees would be recorded as an error of 6.

This figure supports the first two points. The accuracy of prediction has increased greatly over the forty-year period (all the lines trend down and to the right), but error still increases with every day into the future. The mean error three days out in 1971 was about 6 (the bottom line in figure 4.1), and dropped to about 3 by 2019. The difference in error between day 3 and day 7, however, was about 2 in 1999 and was similar, or maybe a little higher, in 2019. All five lines in the graph move in parallel. The Weather Service site also offers accuracy comparisons with four commercial sites. The results are very similar. Given all the talk about the accuracy of weather forecasts, kudos to the Weather Service folks for making this kind of information available.

In addition to this kind of statistical comparison, there is a second approach that asks this question: At what point is a professional, model- and data-intensive forecast no more accurate than using either yesterday's numbers to predict today, or using the long-term average temperature for that place and day of the year? The first method is prediction by persistence, the second is by climatology. Although the concepts are straightforward, it is hard to find data comparing predictions with persistence or climatology, especially for predictions more than a week ahead. I have had my classes do a simple three-month comparison of predicted versus observed weather, using both the U.S. Weather Service and commercial sites, and found that climatology, or long-term averages, beat model predictions after seven to ten days. Persistence never does well.

What about precipitation? Long-term data on the U.S. Weather Service website focuses on the fraction of time that forecasts of more than a certain amount of precipitation are accurate. The "score" for this measure is the extent to which the predicted footprint of precipitation over a certain level overlaps with the measured amount (figure 4.2). The "threat score" varies from 0 (no overlap) to 1 (perfect overlap).

Data on the accuracy of the threat score show the same trend as for temperature: forecasts have been getting better, but skill measured in this way is still less than 50 percent, and it declines with days into the future (figure 4.2). Other figures from the same U.S. Weather Service website show further declines in accuracy four to five and six to seven days out.

One comment on threat score. I think this is a very rigorous and challenging statistic. If rain is predicted where you are and not in a location 300 km away, and the storm veers just a little from its predicted path, then the prediction is wrong in both places, even though the prediction that there would be a storm in the area was right. This highlights how very, very small perturbations to the weather/energy machine can mean big changes in your personal weather.

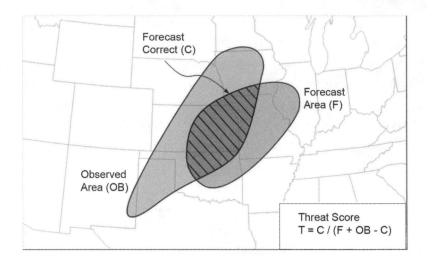

Forecast Correct (C)

Forecast Area (F)

Observed Area (OB)

Threat Score
$$T = C / (F + OB - C)$$

Annual WPC Threat Scores: 2.00 Inches
Day 1 / Day 2 / Day 3

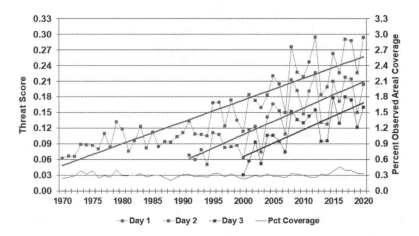

—■— Day 1   —■— Day 2   —■— Day 3   —— Pct Coverage

Figure 4.2. (top) Skill is measured as the degree of overlap between predicted and actual precipitation over a certain amount. (bottom) Predictions have improved over time, but they also decline rapidly with days into the future (the lines from the top are one, two, and three days into the future). Higher threat scores represent more accurate predictions.

Our local concerns are small potatoes compared to the scope and power of the global weather/energy system. Humbling.

What conclusions do some experts draw from data like these, and from their broader experience, about the accuracy of weather forecasts?

A standard meteorology textbook (*Meteorology Today*, by C. Donald Ahrens and Robert Henson) says that forecasts for the next twelve to twenty-four hours are usually quite accurate and that predictions for two to five days out are fairly good, but that beyond seven days accuracy falls off rapidly. They also note that there are problems defining what accuracy means. That was a surprise.

In *The Weather Machine*, Andrew Blum provides valuable insights on how the weather forecasts are produced and talks with many of the scientists involved. In chapter 2 I described the structure of the models and said that those models are updated every twelve hours based on new observations. This twelve-hour update schedule suggests that errors in prediction accumulate fairly rapidly, although Blum quotes one scientist involved in the process as saying that the twelve-hour forecasts are pretty accurate, such that the twelve-hour observational updates result in relatively small corrections.

In *Global Warming: The Complete Briefing*, John Houghton discusses accuracy as reductions in errors in forecasting of atmospheric pressure, reflecting the movement of major air masses, not temperature or precipitation. He presents data showing the same trend as in figure 4.1, that accuracy has improved consistently over the past fifty years but still declines with days into the future. He concludes that three-day forecasts now are as good as two-day forecasts ten years ago. He aligns this with another graph on increases in available computer power in terms of number of calculations possible per unit of time. That one-day improvement in forecasts over ten years coincides with a hundredfold increase in computing power.

Conclusions? Let's start with the positives.

Let's agree that weather forecasting has indeed been getting better consistently over the past five to six decades and that this has been driven by improved observations, including satellite data, and improved computer power. Basically, increased computer power allows the modelers to divide the atmosphere into finer and finer boxes, and the enhanced observations allow better understanding of the initial conditions in each box.

This is really an amazing success story, combining science, technology, and international cooperation among researchers and technicians from around the world. Blum captures this collaboration wonderfully in his descriptions of the open work and social environment he encountered at the European Center for Medium-Range Weather Forecasting (ECMWF). The free flow of ideas and the constant testing of new approaches is a scientist's dream, perhaps something of a modern-day Stockholm Physics Society.

And the importance of accurate short-term forecasts, especially "now-casting" as it is called, focusing on the next several hours, cannot be overemphasized. Being prepared for extreme events is crucial and is reflected in the constantly updated reports on the progress of storms and when and where we should be ready to take cover.

But the hard reality is that no matter how much short-term forecasts improve, accuracy will still decline for both temperature and precipitation as we look further into the future, with climatological averages becoming just as accurate as those modeled forecasts beyond a certain horizon.

Which brings us back to question 3—Why is this so hard? Why does it take such a herculean effort over ten years to improve the accuracy of forecasts by one day? And why does accuracy fade so quickly and consistently as we look just a few days into the future?

Here is where we run into what some meteorologists call "the wall," which can be summed up in a single word—chaos!

Drawn from the ancient Greek, meaning abyss or total void, chaos was what came before all. The concept has taken many forms over the past three thousand years, but synonyms include disorder, confusion, mayhem, havoc, turmoil, anarchy. You can follow the concept through literature and religion, even to Agent Smart and Kaos of old-time TV fame, but let's restrict it here to meteorology, where it has a very particular meaning.

And maybe the best way to give a feeling for the concept is to repeat the classic story of Edward Lorenz. In 1961, the story goes, Lorenz was doing weather research using a simple computer model. He wanted to repeat a previous simulation and reentered what his printout had shown as being the state of the system variables at a particular point in time. He restarted the run with those precise (he thought) values, but the model produced an entirely different outcome. It turned out that he had reentered those variables with only three decimal places of precision instead of the six the computer was using. These very small differences in parameters (less than one part in a thousand) led to a very different prediction.

Stripped down to its essentials, chaotic systems (either physical or mathematical) are those that show extreme sensitivity to initial conditions. This results in part because changes in one part of the system can cause even larger changes in another (technical term: relationships are nonlinear), and those larger changes can cycle back to change the first part (technical term: feedback).

Metaphors and examples of chaos abound. Chaos has been characterized as the butterfly effect from the analogy offered that the flit of a butterfly wing in one part of the world can cause a tornado halfway round the globe. Or as Lorenz once wrote, the flap of a seagull's wings could alter the weather forever. Of course, given that the weather system is chaotic, you could never prove that butterfly did not cause that tornado! And we have only one Earth, so you can't repeat the seagull experiment!

How about a demonstration? You can find lots of videos of the classic double pendulum, and any number of graphs of systems of equations that demonstrate chaos, but, beautiful as some of them are, they always seem a little abstract to me. So let's try this.

Hold a paper facial tissue, or really anything that is light and flat, flexible and wide, at eye level, then let it drop. Watch how it falls and where it lands. Pick it up and start it in *exactly* the same place and let it drop in *exactly* the same way. Did it follow the same path and land in the same place? Probably not. No matter how hard you try, the object will never start in *exactly* the same place—slightly different initial conditions lead to very different outcomes.

If, however, you run this experiment hundreds of times, you might notice a pattern. The tissue ends up in different places, but there are limits. It doesn't fly out the window or through the door. If you could plot the landing place each time, you would see that there are boundaries and probabilities. Essentially, there is a pattern to the outcomes. We will keep this in mind when we start to talk about climate instead of weather—individual events (weather) are hard to predict, but long-term outcomes (climate) follow a predictable pattern.

How about a demonstration of the existence of the kind of fine-scale patterns of airflow that determine overall atmospheric dynamics? Just how small-scale are these patterns that make this a chaotic system?

For this example, we go to the Harvard Forest in Petersham, Massachusetts, just about the geographic center of the state. The Harvard Forest is the site of the longest-running continuous measurements of the carbon balance (the difference between carbon gain by photosynthesis and carbon loss by respiration and decomposition) of a forest.

The centerpiece of this measurement scheme is a metal-frame tower more than 90 feet tall (figure 4.3), hung with all kinds of instruments to measure the environment within the canopy and just above

Figure 4.3. The Environmental Monitoring System tower at the Harvard Forest in Petersham, Massachusetts.

the tops of the mature oak and maple trees. Let's start at the bottom of this tower and work our way up. (As you can see, you would have to be an experienced, and insured, climber to actually do what we are about to do virtually.)

It is a warm and sunny summer day. At ground level you don't feel any air movement at all. It is pretty dark and very humid. All those leaves are taking in carbon dioxide (photosynthesis) and releasing water vapor (transpiration, the unavoidable loss of water during photosynthesis—the reason plants need so much water). Might feel uncomfortable, and you might wish there was a little breeze.

So you start up the tower (virtually). As you go up, you see more light as more and more of the leaves are below you, instead of between you and the sun. The air feels a little fresher, and there is a little bit of air movement. These changes continue all the way to the top, where you emerge into full sunlight and begin to experience weather as it might be recorded by a weather station, only marginally influenced by the forest you just climbed through.

With one exception. If there is any breeze at all, you can feel that the wind shifts minute to minute, sometimes moving up through the canopy, sometimes down into the canopy.

Why have we made this trip? To see two instruments at the top. One measures those changes in air movement you can feel, and the other measures the concentration of carbon dioxide in the packets of air moving up and down across the top of the canopy.

You are at the top of a tower measuring the impact of this forest on the carbon dioxide concentration of the atmosphere. It is called an eddy covariance tower, and the reason for the name is the reason we are here. Those quick changes in air movement (eddies) carry carbon dioxide either into or out of the forest. On a bright sunny day, there is less carbon dioxide in the packets of air moving up than in those moving down. Photosynthesis is removing carbon dioxide from the atmosphere. The reverse happens at night when photosynthesis is shut down but respiration and decomposition continue. The covariance, or correlation, of the direction and speed of air movement with the concentration of carbon dioxide gives you the current impact of the forest on the atmosphere. Carbon gained or carbon lost.

A question that brings us back to chaos and the problem of modeling a finely structured and rapidly changing atmosphere is this: How often do you have to measure wind speed and direction, as well as carbon dioxide concentration, to get accurate results? The answer: six times per second!

The forest canopy presents what is called a "rough" surface to the atmosphere, so that even light breezes passing over (and into) the canopy cause turbulence and the formation of eddies (figure 4.4). You

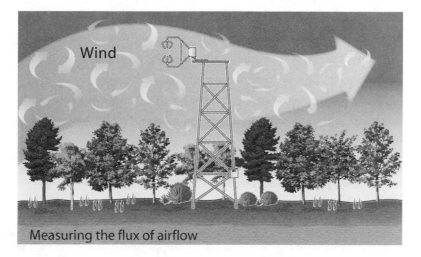

Wind

Measuring the flux of airflow

Figure 4.4. Wind passing over the "rough" surface of a forest canopy creates turbulence resulting in eddies (circular arrows) of air movement. Wind speed and gas concentrations are measured at the top of the tower six times per second to capture the chaotic movement of these eddies.

sensed this in the difference in air movement from the forest floor to the canopy top.

Turbulence is just one process that sets up rapid changes in the structure of the atmosphere and has strong feedbacks through energy exchange, condensation, evaporation, cloud formation, and other processes that all make the atmosphere, at least at the coarse scale we can measure it, a chaotic system.

The science, mathematics, and mythology of chaos theory fill volumes. What does it mean for weather prediction? Very simply, it means that unless those predictive models can start with a perfect three-dimensional description of all the variables in every equation in the model (the correct current values for each of the hundreds of thousands of boxes in the model), predictive success will diminish with time. This is just what we have seen in the earlier figures here. Forecasts have improved demonstrably across the decades, as the observations, computing power, and data assimilation techniques have improved, but the trend in accuracy as you predict further into the future is always the same—the further out you go, the less accurate the prediction.

Can we expect continued improvement in forecast accuracy, or is there a maximum horizon beyond which we will never be able to see in terms of daily weather prediction? John Houghton tells us that even with a perfect model and perfect information, accurate forecasts might be pushed out only from six days to twenty.

Sam Kean's *Caesar's Last Breath* is an enjoyable history of the discovery and uses of gases, from nitrogen (78 percent of the atmosphere) to the chlorofluorocarbons that attack the ozone layer at concentrations of parts per trillion (yes, one in 1,000,000,000,000). Kean offers two impressions that fit here. The first is that the influence of chaos on weather is the first time in his book that "our trusty old gas laws come up short . . . the rush of gases across a spinning planet gets so frenzied that nice, clean simple volume-temperature-pressure relationships can't keep up. . . . Chaos always wins."

The second impression deals with the turbulence that we have just encountered at the top of the Harvard Forest tower, and that permeates all levels of the atmosphere. On this, Kean notes that there is an unclaimed prize of a million dollars out there for anyone who can solve the equations that define turbulence. He also relates that the famous physicist Werner Heisenberg said on his deathbed he intended to ask God for the solutions to two concepts, general relativity and turbulence, and then said, "He may have an answer to the first question." Perhaps apocryphal, but you get the idea.

So we are in good company accepting that the global weather system may escape perfect prediction for the foreseeable future. Still, the complex weather measurement and modeling system that the global meteorological community has created is incredibly valuable for making those short-term predictions that save lives and limit property damage. Continuing research and improvements in measurement may extend somewhat the predictive time horizon for this invaluable service.

Which bring us to the next logical question. If we can't predict weather more than a couple of days out, how can we possibly predict changes in climate over the next century? To answer that question, we need to talk about the differences between weather and climate, and what controls each.

As I said in the discussion about models in chapter 2, each level of understanding exhibits relationships with different controlling variables. Houghton summarizes this by saying that for seasonal (or longer) forecasts, the initial condition of the atmosphere is not the critical piece of information; rather, what controls our long-term climate future is conditions on the surface (for example, ocean temperatures and currents, patterns of ice formation and melt) and how these change over time in response to cumulative changes in the weather.

At even longer time scales, factors outside the climate system, like subtle changes in the position of the Earth relative to the sun, or changes in the position of continents and periods of volcanic activity, can drive change.

For now, let's be thankful that we can have some idea of what the weather/energy machine has in store for us for the next three to five days. Given the data on forecast accuracy like those in figures 4.1 and 4.2, it seems reasonable that the U.S. Weather Service limits its predictions of temperature and type of weather to seven days into the future, and precipitation amounts to three days. Although I think the Weather Service has it right in this, I am sure that I will keep looking at those ten-day forecasts on the commercial sites as well, just to see how much they change as each actual day approaches. Stay tuned!

5

El Niño Is Only the Beginning:
Major Climate Oscillations

What is the difference between weather and climate? It's not too hard to give an intuitive answer.

Weather is what happens today. Climate is longer-term.

Weather is what is predicted over the next seven to ten days by the weather models already presented. It covers the dynamics of the atmosphere that can be measured and tracked using basic physics and the complex observation systems the world has constructed.

Climate is often summarized as the change in average temperature and precipitation over the course of a year based on many years of measurements. In a sense, climate is more predictable than weather. You could probably predict the average temperature for any month next year more accurately than you could predict the high temperature for a day three weeks from now.

But when weather diverges from climatic expectations, that can be big news. Was this a snowy winter? An unusually dry summer? And the big question is, Why?

Behind that big question is this one: Was this summer unusually dry, or is this the new normal? How do we distinguish between this was an unusually hot summer and it is time to install air conditioning because summers will be getting hotter?

There are locations in the Earth's climate system that exert special influence on climate and weather over large areas, and that show cyclical or repeating changes in state or condition over periods of weeks to years. These are called oscillations, and they can cause major shifts from normal weather, making it harder to detect progressive or directional change in crucial numbers like globally averaged temperature, making it harder to answer that question about air conditioning.

Here is one way that I envision oscillations and climate change. The first day of middle school, or maybe even the orientation week at college, or even a team-building exercise at work, can involve a number of community-building games. One that provides a possible analogy for oscillations and climate change is the parachute game. A large circle of light parachute material is placed on the ground and is surrounded by ten to twenty kids (or adults) evenly spaced around the perimeter. Each person grabs the parachute at their location and then lifts and lowers their hold either at will or under direction (you can find lots of videos of this game online). The energy and action of the holders result in waves of rising and falling fabric that travel across the circle in a never-ending pattern of visible hills and valleys. Great fun.

Let's say that the average distance between the parachute and the ground represents global average temperature, and the holders are the oscillations. Given all those constantly changing hills and valleys, how can you tell if the average distance between the ground and the fabric (global temperature) is changing? A scientist might ask, How can you distinguish the signal (change in average height) from the noise (the fluctuations)?

You are going to notice an approach here to oscillations and climate very different from the one we took with weather in the last three chapters. This is our first step away from the defined equations that can be coded into the data-hungry models used to predict weather. There are no first-principle equations so far that can predict the timing and intensity of these oscillations. Relationships tend to be statistical, with underlying physical principles supporting the statistics. Some of the latest approaches actually use artificial intelligence in an attempt to identify trends and predictive relationships in the wealth of data collected that defines these oscillations.

Several important oscillations have been identified and can be described in detail (three of these will be presented here). We have statistical evidence that some of the major oscillations can affect weather at some distance from where the oscillation occurs ("teleconnections" will be part of the presentations), and some understanding of why. It is safe to say that we cannot yet predict the timing and intensity of different phases of these oscillations more than a few months out, or how one oscillation might interact with another.

The Wikipedia page entitled "Climate Variability and Change" identifies ten major oscillations. Other lists can be found, as well as different descriptions, or even names, of each. What this says to me is that oscillations represent one of the most interesting and active areas of climate research. There is much to learn about oscillations and how each impacts the climate system in what can also be seen as an ever-changing pattern—like the fabric in that parachute game.

Let's look at three oscillations that occur at very different time scales and affect different parts of the climate system: El Niño, the polar vortex, and the Atlantic meridional overturning circulation (AMOC). You might be familiar with the first two, as they have been associated with some of the most extreme weather events in the northern hemisphere over the past several decades, but the third could be the most important for long-term climate change.

Remember that we are trying to discern directional, long-term changes in climate by understanding and, if possible, eliminating the influence of these cyclical oscillations on long-term trends.

El Niño is a tale of three seasons.

Season 1: The winter of 1982–1983 was one of the worst in history for storm damage along the Southern California coast. A previous

storm cycle in 1972–1973 had triggered some interest in the phenomenon known as El Niño, but according to climate historian Brian Fagan in *Floods, Famines and Emperors*, scientists focusing on El Niño
worked in relative obscurity for a decade. Then came the El Niño of
1982–1983. Wave after wave of storms caused severe coastal flooding in
Southern California and billions of dollars in damage. Streets were
flooded. Major piers were washed away. El Niño had arrived as a topic
of public awareness, as shown by the number of magazine and newspaper cartoons! But maybe it still seemed a little hard to believe that a
slight warming of ocean waters in the equatorial Pacific thousands of
kilometers away could trigger such devastation.

Season 2: The winter of 1997–1998 was shaping up to be another
big one for El Niño. With fifteen years of research and better monitoring of ocean temperatures, as well as historical reconstructions going
back centuries, El Niño was accepted not just by the scientific community but by decision-makers as well. Central Pacific sea surface
temperatures were rising by early April of 1997, and in June NOAA
(National Oceanic and Atmospheric Administration) issued seasonal
climate forecasts that, in hindsight, turned out to be quite accurate. As
a result of the warnings, agencies took steps to prepare for and miti-
gate the expected storms. Damage was still severe, estimated at $1.1
billion, but that was half of the $2.2 billion estimate for 1982–1983.

I was visiting family in Los Angeles in January 1998 and saw first-
hand some of those preparations. Along the coast, beach sand had been
piled up 10 feet and more in front of near-shore buildings in a successful effort to limit damage from expected storm surge. Good science led
to coordinated action by public agencies to limit the damage.

Season 3: The biggest El Niño event of all time took shape in the
winter of 2015–2016. With better early warning systems in place, predictions months in advance were again accurate, in terms of sea surface temperatures in the Pacific. California was at this time in the

throes of a once-in-a-century drought. Three years of record low rainfall in the agriculturally rich Central Valley and nearly absent snow cover in the Sierra Nevada Mountains had produced major water shortages. Crop production shifted to other states. Severe water restrictions were in place in the Los Angeles area.

The historical correlation between El Niño and rainfall in California foreshadowed an end to the drought—one weather catastrophe to be ended by another. In my environmental science class in the fall of 2015 I boldly and confidently predicted a deluge of rain for the parched region. After all, isn't this what El Niño had always meant—much higher than average rainfall in California?

The winter of 2015–2016 came and went, and the California drought continued. No increase in rain for Los Angeles; in fact rainfall was slightly less than normal. It was only in the following year, after the El Niño was over, that the drought was broken.

So what happened? Let's step back and describe this phenomenon called El Niño, how it is measured, and how it can affect weather in California. After we're convinced that we know all about El Niño, let's see why my bold prediction in 2015 was so wrong.

The scientific term for El Niño is El Niño Southern Oscillation, or ENSO. El Niño derives from the original name given to the phenomenon by Peruvian fishermen who recognized the periodic reductions in the rich fisheries off their coasts around Christmas time (El Niño meaning the boy child). Fagan's history of El Niño captures historical records identifying El Niño events going back hundreds of years, and geologic evidence can trace this recurring cycle back thousands of years.

Uncovering what "oscillation" means in this name introduces us to Sir Gilbert Walker and the problem of the Indian Monsoon. Walker was not a meteorologist, but he was trained as a mathematician and physicist. In 1903 he joined the British Foreign Service in India and was assigned to the office that was charged with predicting the intensity or failure of the monsoon season, key to forestalling widespread episodic famine.

Walker's combination of skills and scientific zeal led him to accumulate all the data on air pressure, temperature, and precipitation that he could find in the early 1900s, and he began to establish regional patterns in these data. In particular, he noticed that there were oscillations in the measurements and that these tended to correlate with weather patterns as well. He named two major oscillations: the Southern Oscillation (now called ENSO) and the North Atlantic Oscillation.

For El Niño, these differences are part of an integrated atmosphere-ocean pattern, in turn part of the much larger system redistributing solar energy throughout the climate system.

Remember the Hadley cells presented in chapter 3? Heated air at the equator rises to the upper atmosphere and moves toward the poles. Completing the Hadley cycle, air descends around 30 degrees north and south (although we saw how wildly this boundary can change when we talked about the jet streams) and then moves along the surface back toward the equator. The Coriolis effect deflects these winds to the southwest in the northern hemisphere and to the northwest in the southern hemisphere.

These winds from north and south meet near the equator in what is termed the Intertropical Convergence Zone (ITCZ; figure 3.2 shows this feature)—a good name, as it defines the convergence of these two air streams. This convergence is captured in descriptions of the trade winds in any good seafaring novel or history—a relatively constant flow of surface air from east to west from Peru to north of Australia.

It is interesting to me that the ITCZ results from the convergence of winds from north and south, and is not determined by the physical location of the equator. Rather, it responds to the uneven distribution of solar energy across the globe, so the ITCZ moves north during the northern hemisphere summer and south in the winter as the angle of orientation between the Earth's axis and the sun completes its annual cycle.

Winds and ocean currents are inextricably linked. The flow of air from east to west creates a parallel ocean current, so seawater also travels east to west until it tends to pile up at the Australian end. The surface of the ocean might be two-thirds of a meter higher at the western end.

Responding to changes in pressure, the water sinks, and you can guess what happens next. To complete the cycle, the water flows back to the east at depth, cooling as it travels and gathering nutrients as well. When this counterflow is blocked by South America, it upwells to the surface, bringing cold, nutrient rich waters to the coast of Peru. A rich anchovy harvest follows.

In honor of Sir Gilbert Walker, this cycle of air and seawater is called the Walker circulation (figure 5.1).

This is the normal or neutral condition. An El Niño event occurs when this cycle slows down or stops. The trade winds falter and sometimes reverse. Without the winds, the ocean circulation also shuts down, and the cool seawater at depth stays there. Surface ocean temperatures off the coast of Peru can climb by as much as 5°C over a very short time, and the anchovies disappear.

You may have heard of another term, "La Niña," that has entered the ENSO lexicon. Meaning "the girl child," this expresses the condition opposite to El Niño. During a La Niña event, the trade winds are stronger, as are the ocean currents; the entire cycle speeds up.

As the system shifts from one state (El Niño) to another (La Niña), every other measurable climate variable changes as well. Wind speeds and directions shift, air pressure changes, cloud formation and

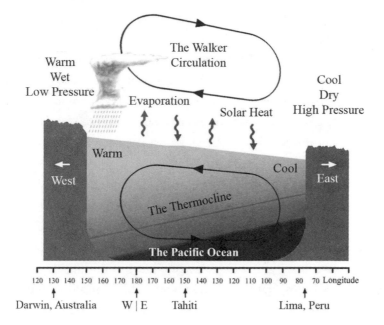

Figure 5.1. The Walker circulation, showing the normal cycles of movement in both the atmosphere and the ocean in the equatorial Pacific.

precipitation shift. Most notably, a strong El Niño usually means heavy rains and floods in Peru but drought and fires in Australia, while La Niña means rainfall for Australia and even drier than normal conditions in Peru. Impacts on Indian monsoons and even the formation of hurricanes in the Atlantic are attributed to El Niño, although research in these areas is continuing, especially on the interactions among ENSO and other oscillations.

Given that everything meteorological changes, how does the climate community monitor and summarize the state of this oscillating part of the climate system? Monitoring is intensive. Hundreds of moored and floating buoys collect continuous data on all the basic weather data at the surface, as well as temperature, flow rate, salinity, and other characteristics at several intervals all the way to a depth of

2,000 m. Sea surface temperature and even elevation are continuously monitored by satellite.

The complexity and detail of all of these measurements can make it hard to present the current state of the system to the public. For this reason, NOAA uses a simple index to convey this system, to record its history, and to predict its future. That index is the sea surface temperature anomaly in the central equatorial Pacific, a region known as Niño 3.4.

Why sea surface temperature, and why anomaly? Many of the atmospheric responses to El Niño change rapidly and over short distances, as our discussion of weather prediction highlighted. Capturing these would be difficult, and results would be transitory. Ocean temperature, however, cannot change minute to minute, so changes in sea surface temperature provide a more stable indicator of the state of the system.

Anomalies are the currency of climate change. If I told you that the average surface temperature in the central Pacific went from 30°C to 32°C, there would be no context to those numbers. Are they high or low? An anomaly is the difference between a current measurement and the long-term average for that time and place. The Niño 3.4 index is reported as differences from these long-term averages, or sea surface temperature (SST) anomalies.

Images of SST anomalies present the spatial state of the ENSO system in graphic detail (figure 5.2). During a strong El Niño, the SST anomaly on a given day will be as much as plus 4°C. For a strong La Niña, this index might be as much as minus 4°C.

To capture seasonal trends, NOAA averages these values over a running three-month period, reported monthly (figure 5.3). Could this be any more different than the ten-minute cycle of the weather models? This index changes slowly, and NOAA uses the arbitrary cutoff of plus or minus 0.5°C to determine if we are in an El Niño, neu-

El Niño   La Niña

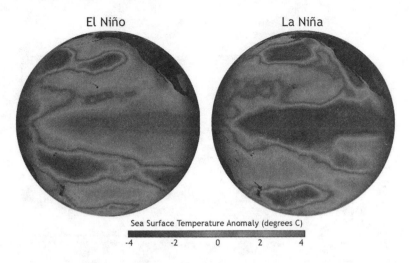

Sea Surface Temperature Anomaly (degrees C)

-4   -2   0   2   4

Figure 5.2. Images of sea surface temperature anomalies in the central Pacific Ocean during an El Niño event (left) and a La Niña event (right). The darker area at the equator in the La Niña image represents colder than average temperatures. The light gray area at the equator in the El Niño image represents warmer than average temperatures. In these images, Peru would be at the right edge and Australia to the left.

tral, or La Niña state. This index has been reconstructed back to 1950. You can see that the four strongest El Niño events (those reaching plus 2°C for this three-month average) are the ones we have discussed: 1972–1973, 1982–1983, 1998–1999, and 2015–2016.

There are sophisticated ways to describe the frequency or return interval of El Niño events of different intensity, but we can do this in an intuitive way. There are about eighteen times in figure 5.3 that the index reaches 0.5 or above. Since the record covers seventy years, the frequency of El Niño events would be (70/18), or every 3.9 years. There have been only four times when the index rose above 2.0, so the return interval for these big ones is (70/4), or 17.5 years.

It might be counterintuitive to think that a shift of a couple of de-grees in sea surface temperature in the central Pacific can wreak havoc

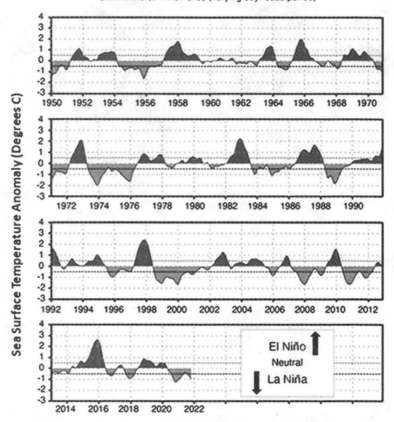

Figure 5.3. The Oceanic Niño 3.4 index. Each monthly value is the average of three months of sea surface temperature anomalies. A value greater than +0.5 is used to identify an El Niño event. One more negative than –0.5 signifies a La Niña event.

on the California coast. And this brings us to teleconnections and back to Sir Gilbert Walker, who is given credit as the first to note that the Southern Oscillation that we now know as ENSO and the Walker circulation could impact the monsoons in India. "Teleconnection" (same root as television and telephone) just means that a change in the climate system in one location alters the weather in a distant location. While a teleconnection might first be suggested just by looking at the statistics, ideally there is an identifiable physical reason for that relationship.

Analyzing global weather and climate data, we find that the strongest relationships for El Niño are with winter rainfall in California. For Los Angeles, total winter rainfall averages about 50 percent higher than normal during El Niño years, and can be lower than normal in La Niña years.

Different descriptions of the effect of El Niño anomalies on climate and weather in different parts of the world can be found, and that suggests to me that this is still an unsettled area of climate science. We know what El Niño is and we can monitor the state of the system in fine detail. Warming of the ocean in the central Pacific releases immense amounts of heat and associated humidity into the atmosphere, which we would expect to ramify at least regionally. It can also affect globally averaged temperatures.

The primary change in the climate system attributed to El Niño is a shifting of the normal path of the jet stream over North America in winter. Rather than traveling north across Canada and down through the northeastern United States, bringing cold and snowy weather, the jet stream shifts south, crossing the southern United States, bringing increased rainfall, especially to California. A La Niña year can be colder and snowier in the Northeast, drier in California.

So this all did seem like settled science to me in the fall of 2015. The statistics were strong, and the mechanism made sense. I therefore confidently predicted to my class of first-year students that the cataclysmic drought in California would end in the El Niño winter of 2016. But it didn't. What happened?

At the same time that the El Niño was forming, a stationary pool of anomalously warm ocean water, popularly christened the Blob, was also forming off the west coast of North America. Running from Northern California up to Alaska and hundreds of kilometers out to sea, the Blob was revealed as sea surface temperatures as much as 3°C above average (the anomaly). This had some disastrous effects on marine life, from plankton up to whales and seabirds, and has fostered a new field of research into "ocean heat waves." It also bred another round of cartoons—the Blob versus El Niño!

The Blob is given credit for altering the usual diversion of the jet stream to the south during an El Niño event and blocking the usual increase in rainfall along the California coast. The drought in California actually ended in the winter of 2016–2017 during a mild La Niña.

So there are some unknowns regarding El Niño. Predicting its occurrence is one of those. The U.S. Weather Service site for El Niño and La Niña includes monthly reports on status and measurements, and also predictions for the upcoming six months made by several different models. The predictions cover a wide range, but the average of them all bears some resemblance to what actually happens. And kudos to the Weather Service again for publishing not only predictions but also including in the same graphs how accurate previous predictions proved to be.

So we want accurate predictions about oscillations like El Niño because they can cause major if temporary shifts in the climate system that can mask progressive change and cause serious damage, but our understanding is still imperfect. We have come a long way from Walker's hand calculations of rough patterns in atmospheric pressure, but additional discoveries await. And the Blob serves as a cautionary tale saying that surprises await in terms of the science of climate oscillations like El Niño.

Let's look at another oscillation—one that operates in a much shorter time frame.

The winters of 2013–2015 included some record-breaking cold spells in the northeastern United States and eastern Canada, and suddenly the polar vortex was the weather topic (and cartoon topic) of the day. Average temperatures for all of New England over those three winters were 5°C below normal for February, and 2°C below normal for March. Snowfall records were set throughout the region. One congressman famously gathered a snowball from the lawn outside the Capitol building in Washington and brought it into the chamber, claiming this showed that global warming could not be real.

Ironically, it is not the polar vortex that caused those outbreaks of cold weather but the breakdown of that vortex.

Let's go back to Hadley, Ferrel, and Polar cells again. The polar jet stream created by the turbulent interaction of the Ferrel and Polar cells roars around the northern hemisphere at up to or more than 150 km per hour, driven by differences in temperature, pressure, and height of the two interacting cells (see figures 3.3 and 3.4). The strength of this jet is related to the difference in temperature between the cells. That difference is greater in the winter, when the difference in sun angle between the two air masses is greatest, so this jet roars a little louder in the winter.

Under normal conditions, this jet strikes a straight line between the cells isolating the coldest arctic air to the north from the more moderate temperate-zone air to the south. This is the polar vortex. Those persistent and extreme cold weather outbreaks in the northeastern United States and eastern Canada resulted from the breakdown of this pattern of isolating wind flow allowing supercold arctic air to escape to the south. At the same time a blocking pattern in the atmosphere kept this southward flow of cold air locked in place over the northeastern United States, resulting in not just a cold snap but also a record-cold winter.

How does this happen? The vortex tends to be stronger and more continuous in the upper atmosphere (the stratosphere), trapping cold arctic air over the north pole. Occasionally the warmer air in the lower atmosphere (the troposphere) kicks up into the stratosphere, leading to sudden stratospheric warming. This reduces the temperature gradient between the air masses to the north and south of the jet stream between the Ferrel and Polar cells, weakening the jet stream, causing it to wobble or break down and increasing the chance that some of that frigid arctic air will escape to the south.

This is what happened in the winters of 2013–2015. At one point, NASA reported a 25°C increase in temperature in the stratosphere over the Arctic within a single week!

So another irony here is that it was a warming of the upper atmosphere over the Arctic, part of a general warming of the Earth's climate system, that led to this outbreak of supercold air. That snowball in Congress is just a local cold expression of a global warming pattern.

Another impact of the blocking pattern was that outbreaks of warm air in other parts of the world during those three winters tended to remain in place as well. While the eastern United States was freezing, the western half of the country was experiencing unusual warmth. Globally, those three years were among the warmest ever recorded (as is true for every year in this century). Warming winters in parts of the United States don't create the kinds of disaster headlines that frigid weather and blizzards do, but you may have noticed the overlap in time period for both the severe drought in California and the severe cold in the northeastern states.

Can we track and predict these kinds of outbreaks? Are there indices such as the Niño 3.4 temperature anomaly that capture changes in this part of the climate system? Introducing the Arctic Oscillation.

The Arctic Oscillation index tracks changes in air pressure within the Polar cell. When the index is positive, lower than normal pressure keeps the jet stream tightly circling the pole. A negative index means higher pressure and increased likelihood of more wobble (to use the technical term), and more cold-weather outbreaks. The strongest teleconnection is the one described here with cold-weather outbreaks to the south, but the locations of those cold outbreaks, and the corresponding warm areas at the same latitude, are hard to predict.

Predictions of future values for the Arctic Oscillation index are reported on the U.S. Weather Service Climate Prediction Center site. It is interesting that, as an atmospheric phenomenon, the index is reported and predicted daily, but technical papers relating these to outcomes also use monthly averaged values. Weather or climate, or both? For the daily values, the Weather Service site shows what we have come to expect from weather predictions: good for the next couple of days but declining over the reported fourteen-day time line.

The Weather Service publishes both predictions for the future and the accuracy of previous predictions for this oscillation on the same graph. This is the only site I can find that does this as a matter of general practice. The Climate Prediction Center is well worth a visit.

As with other probes into the climate system, we are good at measuring the Arctic Oscillation index and relating it to outcomes but less good at prediction. In a sense, we know who is shaking these parts of the weather parachute (to return to our opening analogy), but we can't predict who is going to shake their part of the system when, or how hard.

It is also clear that different movers and shakers of the parachute differ in strength, in how hard they shake the system, and over what period of time. The Arctic Oscillation is atmospheric, so it shifts frequently and relates primarily to short-term or possibly seasonal weather. El Niño is measured using an ocean-based index reported monthly and averaged over a three-month period, and can force atmospheric changes summarized seasonally or annually.

Now we come to the biggest shaker of the three on our list: the Atlantic meridional overturning circulation (or AMOC).

This may be more familiar to you as the Gulf Stream. Those intrepid seafarers in the age of sail knew all about the Gulf Stream, a

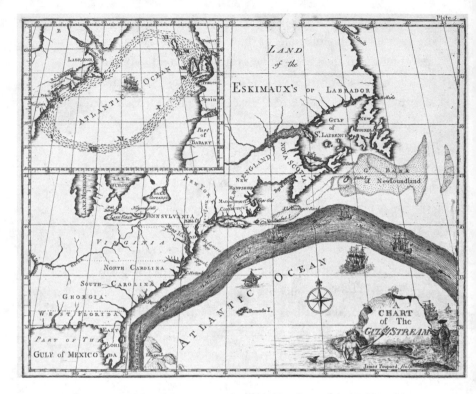

*Figure 5.4. (above) Benjamin Franklin's first chart of the Gulf Stream;
(facing page) a more modern and accurate rendering.*

wide band of warmer than expected ocean water flowing southwest to
northeast from the Caribbean toward northern Europe.

Smithsonian magazine gives credit to Benjamin Franklin for being
the first to chart the Gulf Stream (figure 5.4, above), following a lead
from a cousin who had experienced the flow as the captain of a mer-
chant ship. A more modern and more accurate rendering of its path
(figure 5.4, facing page) shows that this stream actually reaches much
farther north, into the North Sea between Scandinavia and Iceland.

The Gulf Stream transports unfathomable amounts of tropical
warmth from the equatorial regions to northern Europe, with some

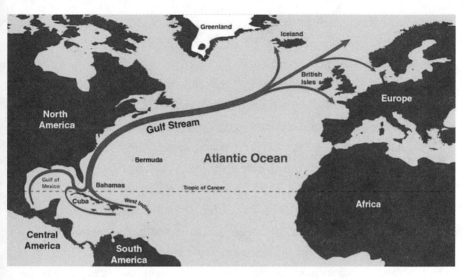

influence on the East Coast of North America as well. It also drives, and to some extent directs, the paths of hurricanes, which are likewise an important part of the transfer of tropical heat from south to north (in the northern hemisphere).

It would be hard to overstate the importance of the Gulf Stream in the modern world. The climate in Oslo, Norway, at about 60 degrees north latitude, is vastly more temperate than that in Anchorage, Alaska, at about the same latitude. Winters in Ireland and England can be cloudy and wet, but snow is rare, unlike winters in Calgary or Winnipeg, which are actually 2 degrees farther south. You won't find the kinds of palm trees you can see in the south of England and Ireland much north of the Carolinas in the United States. Agriculture, economics, cultural history, and just about everything else about northern Europe, and the other parts of the world historically influenced by that region, derive and depend in large part on the flow of the Gulf Stream.

We can ask what causes this mighty flow of energy, but we could also ask the same question in another way. If the trade winds that drive El Niño result from wind patterns that should be repeated over the Atlantic, why is there no Walker circulation there? The short answer is: the American continents.

There is a consistent ocean current flow from the coast of Africa toward Brazil and the Caribbean. The North and South Equatorial currents meet near the Intertropical Convergence Zone in the Atlantic, just as in the Pacific, and carry water from east to west. But then these joined currents run up against South America. So while the equatorial current in the Pacific is unhindered for nearly 15,000 km, and then flows on into the Indian Ocean, the equivalent Atlantic current meets this impenetrable barrier and shifts north. At that point we change the name to the Gulf Stream, but it is all one flow.

And it is a critical piece of the total global energy balance as well as global-scale ocean flows. Only in recent decades has it become clear that all the major ocean currents in the world are linked in one continuous—you guessed it—cycle. The full name for this cycle is the thermohaline circulation, derived from the understanding that differences in temperature as well as the concentration of salt drive the system.

The Gulf Stream waters cool and become more salty (through evaporation) as they travel north. After delivering the gift of warmth to northern Europe, the water sinks, as colder, saltier water will do. This is thought to be a key driver of the entire global ocean circulation system.

So monitoring this key rate of flow is central to assessing changes in the global energy redistribution system that drives climate. Basically, if the rate of flow declines, less warmth will be delivered to northern Europe, with potentially disastrous consequences. Other key climatic patterns in nearly every part of the world could shift as well.

We have known of the Gulf Stream for centuries, but the total flow of water in this current is so massive that only recently has a detailed program of measurement been initiated. The program is operated jointly by the United States and United Kingdom and is called RAPID. The index used to monitor AMOC is the averaged rate of flow expressed as Sverdrups. One Sverdrup, eponymously named for its inventor, Harald Sverdrup, is the transport of 1 million cubic meters, or 260 million gallons, of water per second. That is a big number!

The fine-scale data from the RAPID project show large variations in flow rate over the first fifteen years of monitoring, but no discernible trend. Attempts to reconstruct longer-term estimates have suggested either a sixty-to-ninety-year cycle in the system or a rapid recent decline. With a rigorous program of measurement now in place, it may be possible to answer the question of cyclical versus directional change, but as with our parachute example, it may take some time for the long-term signal to emerge above the level of the interannual variation, or noise.

So weather and climate are both driven by differential amounts of received solar energy, interacting at different time scales with geography as described by basic physical principles. How rapidly the different oscillations change varies widely. Purely atmospheric processes, such as the Arctic Oscillation, change daily. The largest-scale oceanic processes, like AMOC, change only over decades. El Niño, involving the interaction of wind with ocean currents, changes at about a five-to-fifteen-year frequency, depending on the intensity.

To assess climate change, the key questions are not only about return intervals of oscillations but also about directional change. Could modification in the Gulf Stream and AMOC create different conditions that would alter the intensity or frequency of ENSO events? Could changes in decadal-scale oscillations like ENSO alter the intensity and frequency of breakdowns in the polar vortex? How do we differentiate between repeating cycles and directional movement to a new climate state? This could be one of the most interesting areas of climate research in the near future.

Three relatively well-known and important oscillations have been presented here, but there are many others under study. The names invoke nearly every place and part of the climate system and operate at different time scales (for example, the North Atlantic oscillation, the Indian dipole oscillation, the North Pacific gyre oscillation, the interdecadal Pacific oscillation). This is an active area of research. We have a lot to learn about who is holding on to that climate parachute, and how hard, and how frequently, they can shake it!

There is another part of the climate system that responds even more slowly than AMOC and the thermohaline circulation system. The encroachment and recession of the huge glaciers and ice caps that have defined the term "ice age" also play a role in responding to and controlling climate. Are there lessons hidden in the remaining massive accumulations of ice, especially in Greenland and Antarctica, that can teach us not only about longer-term climate represented by the ice ages but also just how rapidly our climate can shift from one state to another?

6

What Can Ice Tell Us about Rapid Climate Change?

We have passed through a gray area between weather and climate. The short-term changes in weather described in chapters 2 through 4 can be affected by shifts in oscillations, some of which were described in chapter 5. Extending our reach beyond a few years and into the realm of long-term climate dynamics changes the game entirely. Not that the physical rules of the game are different, but explaining them and making predictions require a shift in focus.

There are two reasons for this shift.

The first is that initial conditions in the atmosphere, which are so important for weather prediction and limit the accuracy of those predictions to a couple of days, become irrelevant at the scale of decades to millennia. Causes and controls for major shifts in the climate system occurring at this time scale need to be sought in parts of the system that change at a similar pace.

For example, this chapter explores how subtle shifts in the orientation of the Earth to the sun affect the distribution of energy received at the surface and contribute to the approximately hundred-thousand-year cycle of advance and retreat of global glaciers that we call the ice ages. But sunlight received is not the only driver of change. Changes in the amount of ice on the Earth's surface can become both a first response to altered solar energy and an additional cause.

The key concept here is feedback. Our time frame now is long enough for feedbacks to become not only possible but central.

What are feedbacks? Technically, a feedback is a response to change that amplifies or diminishes the cause of that change. Amplification is called positive feedback, and diminishment or damping is called negative feedback. Let me make it clear from the start that positive and negative here are not value judgments! Positive feedbacks can be bad, and negative feedbacks can be good, depending on your point of view.

The analogy that I have found to be the clearest way to capture feedback is your home heating and cooling system. When it works well, this is a classically good negative feedback system. You set the thermostat (literally, the temperature constant) at 68°F (20°C). The thermostat has an operational range around that number. Let's say it is plus or minus 2 degrees, meaning that the thermostat should keep the house between 66 and 70. It's cold out, and the house goes down to 66. The thermostat triggers the furnace (or whatever), and heat flows into the house. When the air around the thermostat reaches 70, it shuts off the furnace. If you also have air conditioning, that same thermostat will signal the A/C to crank up when the temperature reaches 70 and to shut off when it reaches 66.

This is a good negative feedback system because the change in the system (temperature in the house) triggers a response (furnace or A/C) that returns the system to its original state or condition (68 degrees).

As a practical joke, let's say that you switch the response wires at a friend's house so that 70 degrees triggers the furnace and 66 triggers the air conditioning. You can imagine what happens then. Once triggered, the house will continue to get hotter or colder until some outside force (like your angry former friend) turns the whole thing off and rewires the system. Interestingly, where the system goes depends on which trigger temperature is reached first. Cold leads to colder, hot leads to hotter. So this is a bad positive feedback.

One characteristic of this kind of feedback is that the changes can accelerate with time. Imagine if the heat put out by the furnace in that second example increases as the house gets warmer. This would be a strong positive feedback, and the system can become unstable very quickly, possibly reaching a critical threshold, like being hot enough to cause other systems in the house to malfunction.

Feedbacks abound in the climate system, and many of them are positive or destabilizing. These operate within the constraints of a few major driving forces that have kept the Earth's climate habitable for living things for billions of years. But those destabilizing positive feedbacks suggest that the Earth's climate system should have been subjected to very large changes in average temperature over the course of its history. Has it?

This chapter focuses on the past million years or so and talks about the ice ages. The next looks back hundreds of millions of years.

There are many stories locked in ice, and ice can be a great way to communicate stories as well. I've spent almost all my professional career in places that were covered by ice during the ice ages, so a great entry point for this topic with my students is to ask them to imagine a glacier more than a kilometer thick covering the spot where they stand today. I'll then send them searching (online of course) to discover when, and how, that happened. In a class dealing with geologic time frames of millions of years, finding that New England was covered in ice as recently as twelve thousand years ago can be very surprising.

Knowing we have no direct observations of climate in these ancient times, how then do we look back that far to see how the climate system has changed and determine what might have caused those changes? Ice can help. Past changes in climate leave signals behind, and ice can contain some of those signals.

The story of the discovery of past ice ages, and the realization that the northern world was once covered with a much greater extent of ice than we now see, has been told many times, and told well. *Ice Ages* by John and Katherine Palmer Imbrie is an older book that tells the story of the discovery, while Tobias Kruger's *Discovering the Ice Ages* updates and extends that story.

The discovery of ages of massive ice expansion across the northern hemisphere is generally associated with Louis Agassiz, a Swiss paleontologist and ichthyologist. He produced a definitive early book on the phenomenon that built on, but perhaps did not fully acknowledge, the contributions of colleagues and parlayed that into a professorship at Harvard and a career and life as a major actor in the development of several important Boston cultural institutions.

But both Kruger and the Imbries include the accounts of woodcutters, farmers, engineers, and others whose direct experiences led them to assume that ice had once been more widely distributed in Europe, well before mainstream academics laid claim to the discovery of this phenomenon. Familiar both with the scattered distribution of huge boulders about their landscapes and with living in proximity to active glaciers, they assumed that only the actions of gigantic masses of moving ice could possibly explain the distribution of these glacial erratics.

The Imbries retell a story of a Swiss engineer-turned-geologist, Jean de Charpentier, who, while on his way to present the radical idea of previous ice ages at a scientific conference, spoke with a woodcutter who recognized that large boulders in his area were made of granite from a distant mountain range and said that only ice could have moved them so far.

Although recognized by those living with the evidence of past glaciers, the scientific community was highly resistant, as the idea contradicted central tenets of the dominant worldview. This is a classic example of the push for, and resistance to, what is called a paradigm shift—the switch from an existing to a wholly different all-inclusive worldview. Chapter 8 looks at three other examples of paradigm shifts, and the processes and time lines involved.

An important step in a paradigm shift occurs when enough curiosity has been aroused by contradictory evidence that professional and amateur researchers across a wide range of disciplines begin to look at their own subject in a different way. Once the possibility of a massive ice age was raised, oceanographers began to see that sea levels had changed dramatically, reflecting changes in the amount of water stored in glaciers worldwide. Geologists and soil scientists determined that regions with deep, silty soils (called loess) could not have been deposited by water but only by strong winds at the foot of the massive glaciers that redistributed wind-blown dust. Accumulations of poorly sorted rocks and gravel were identified as glacial moraines built from geological debris deposited where glacial advance halted and debris carried by those glaciers was dropped at that terminus. Patterns of scratches in bare rock, called striations, showed in which direction the glaciers moved as they scoured those rocks in their passing.

Acceptance of this paradigm shift (the occurrence of ice ages) brought coherence to a set of what previously seemed to be unrelated observations. Accumulated careful reading of all the signs suggested that not only had there been a glacial age, there had been several, with the distribution and extent of the ice cover differing with each advance.

Once the existence of ice ages was accepted, a natural follow-on was: What could possibly have caused such massive changes? Surely major shifts in global climate would have to be involved, but how big, and what could have caused them?

On the way to the best answer, we pass again through the Stockholm Physics Society and the impact of Svante Arrhenius. Intrigued by the idea of ice ages, Arrhenius was also aware of older research demonstrating the ability of carbon dioxide to absorb heat radiation.

Putting these together with presentations to the society by Arvid Högbom on controls and changes in the carbonic acid content of the atmosphere led Arrhenius to propose that those changes had triggered the ice ages. This led to the massive set of calculations described in chapter 1.

When a new concept like the ice ages becomes accepted, a host of hypotheses will emerge as possible explanations. Once the true answer is known, some of these hypotheses will seem absurd, but this is all part of the process. The give-and-take in the scientific debate on a controversial topic is what makes science interesting and maybe even fun. Trying out new ideas on colleagues and registering their astonishment, disbelief, or perhaps even sudden agreement is a rewarding part of the process. Ideally, disagreement leads to alternate, maybe warring or conflicting, hypotheses, which lead to new research and new measurements, which in turn lead to new information and progress in understanding.

From the beginning, one hypothesis was that the ice ages were the result of changes in solar energy reaching the Earth, but the answer did not come from the sun, it came from how the Earth receives that solar gift.

In *Ice Ages*, the Imbries tell in some detail the story of Milutin Milankovitch, an engineer and mathematician who would join Arrhenius, Bjerknes, and Walker among the colossal calculators of the precomputer age. The story finds the young Milankovitch and a poet friend sharing multiple bottles of wine at a café in 1911 and setting their lifelong ambitions. Life up to that point had led him to a Ph.D. and an early career in engineering before accepting an offer to become a professor of applied mathematics at the University of Belgrade.

As the Imbries repeat the story, Milankovitch suggested later that perhaps the wine had something to do with it, but he states his goal as "understanding the infinite!" Refining that goal in more sober times, he limited the scope to explaining the climates of the Earth, Venus, and Mars, and then finally just to building theoretical calculations for predicting changes in the Earth's climate, based on its orbit around, and inclination toward, the sun.

The parallels here with what Arrhenius accomplished are apparent. A bent for theoretical or synthetic thinking—wanting to solve the whole problem—was one. Awareness of previous research was another. As Milankovitch started his calculations, he found that other scientists had already provided the necessary information about variations in orbit and inclination. A final parallel is a penchant for clear quantitative and mathematical reasoning, and the willingness to spend untold hours completing the painstaking calculations required.

Although Arrhenius's isolation during his time calculating the Second Arrhenius Equation was self-imposed, Milankovitch did not have that luxury. Working twenty years later than Arrhenius, he was disrupted by World War I, though the war provided a window of opportunity as well. Arrested by Austro-Hungarian forces in 1914, he was imprisoned for a short time before being released at the behest of a fellow scientist. He spent the rest of the war basically under house arrest in Budapest—but with access to the library of the Hungarian Academy of Sciences.

During this time in isolation, Milankovitch completed a first set of calculations on the interaction between the occurrence of ice ages, variations in the Earth's orbit and angle of inclination toward the sun, and changes in the amount of energy received at different latitudes. The rest of this history is interesting, but let's move on to see how this interaction between the Earth and the sun affects ice ages.

There are three major characteristics of this interaction that affect the distribution of solar energy received and shifts in climate that result.

First, the orbit of the Earth around the sun in not circular but elliptical, or oval-shaped. That ellipse is not constant but is at times

Milankovitch Cycles

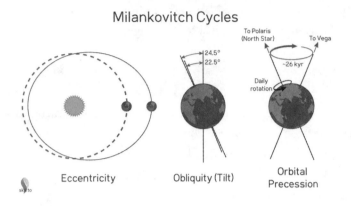

Eccentricity   Obliquity (Tilt)   Orbital Precession

Figure 6.1. Three variations in the orbit of the Earth around the sun causing changes in the distribution of sunlight received, which in turn can trigger the initiation of an ice age.

more round and other times more oval, driven by the gravitational pulls of Jupiter and Saturn (figure 6.1). These changes in the orbital path are called eccentricities and occur on a predictable cycle of about a hundred thousand years. The effect of these noncircular orbits is that summer and winter are, technically, of different length in the different hemispheres, and these shifts reverse as the hundred-thousand-year cycle plays out.

Second, seasons on Earth are driven by the axis of rotation tilting by an average of about 23 degrees. As the Earth orbits the sun, when the north pole is tilted away from the sun, it is winter in the north and summer in the south, and vice versa. This tilt is also not constant but varies between 22.1 and 24.5 degrees (figure 6.1). Again, this variation (also called obliquity) is predictable and describes about a forty-one-thousand-year cycle. More extreme tilt accentuates the effect of seasons on weather and climate.

The third factor is not one process alone but a category of subtle changes in sun-Earth orientation under the general term "precessions." These are well known, but their effects are of lesser magnitude and can be conflicting. The most frequently cited effect is the precession of the equinox, or axial precession, which has also been described by the precise technical term of wobble in the Earth's axis of rotation within the changes caused by the tilt in that axis.

The significance of these subtle shifts in orbit and orientation is they alter the distribution of the sun's energy as received by the Earth. Any process that reduces solar energy received, especially in the far north, which has the landmass to support large ice sheets, has the potential to trigger an ice age. Colder temperatures minimize melting of ice in the summer.

. These are the heart of the calculations that Milankovitch carried out and his relatively accurate prediction of the periodicity of ice ages are recognized in that these are now called Milankovitch cycles. But those changes in calculated solar energy alone are not enough to cause the wide variation in temperature and glaciation that has become evident through decades of geological, oceanographic, and climatological research. Those swings are larger than Milankovitch's calculations would predict.

Which brings us back to feedbacks. The Earth's climate system is too complex, and there are far too many interactions between temperature and response to cover them all here; some remain controversial. Major texts like John Houghton's *Global Warming* describe many of these feedbacks. You can find good science writing related to feedbacks between climate and just about any part of the Earth system.

So let's concentrate just on two we have seen before: ice reflects sunlight, and carbon dioxide provides greenhouse gas warming.

"Albedo" is the technical term for how much light reaching a surface is reflected back into space. It's expressed as a fraction between zero and 1. Zero means that all the light that hits a surface is absorbed (it would look deeply black to us), and 1 would mean all the light is reflected (it would contain the full spectrum of colors and appear white).

Those examples refer only to the kinds of light we can see, but the concept applies to all forms of radiation. Fresh snow has an albedo approaching 1 (really more like 0.85). Water has an albedo value of less than 0.1, meaning that 90 percent of incoming radiation is absorbed, with the energy in that radiation serving mainly to increase the temperature of the water. Forests have a low albedo, while bare ground tends to be more reflective.

So you can see where this is going. If the Milankovitch cycle leads to the growth of glaciers in the far north, those glaciers will reflect sunlight and reduce heat gain. On the other hand, if a warming planet leads to a disappearance of ice (say, as ice cover in the Arctic Ocean), then the water exposed will absorb almost all of whatever solar energy is received and will warm, reducing the chance for new ice formation the following winter. These are both classic positive feedback processes. Getting cold, it will get colder. Getting warm, it will get warmer.

Carbon dioxide can also lead to positive feedback effects. One of the puzzling pieces of information out of the ice cores is that global temperatures across the hundred-thousand-year cycle of the ice ages are correlated with measured carbon dioxide in the atmosphere (figure 6.2). Carbon dioxide concentrations are lower during cold periods and higher during warm periods. Arrhenius would like this and might happily conclude that carbon dioxide is indeed driving the waxing and waning of the ice sheets.

Detailed research on the timing of the carbon dioxide rise and on the processes that control it suggests the opposite—that warming triggers biological and geological processes that lead to an increase in the gas. Warming comes first, and increased carbon dioxide accentuates that warming. Rather than causing the ice age cycles predicted by

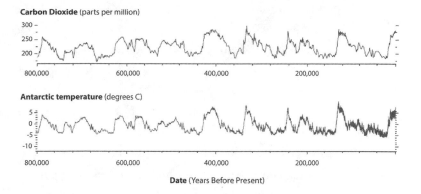

Figure 6.2. Changes in the concentration of carbon dioxide and global temperature anomaly from the Vostok ice core collected in Antarctica.

and driven by changes in orbital characteristics, carbon dioxide appears to increase the response to those cycles.

So we have two major positive feedbacks in play that would seem to drive the global climate system in one of two ways—toward a carbon-dioxide-rich superwarm planet or an ice-covered (high-albedo) snowball. I deal with both of these possibilities in chapter 6. What appears to happen with the cyclic ice ages on Earth is that the orbital effects calculated by Milankovitch, and many others since, eventually override these feedbacks and reset the system back into the larger, radiation-driven cycles. In a sense, your former friend just installed a more powerful thermostat/furnace/air conditioning system that can override your rewiring prank!

Let's wrap up this chapter with an adventure story or two, and a major discovery related to climate change.

Not all branches of climate science offer opportunities for adventure as well as discovery, but glaciology is one that does. In *Thin Ice*, Mark Bowen combines world-class mountain climbing expertise with a degree in physics to explore the rigors of collecting ice from the rapidly disappearing glaciers high on tropical mountains by traveling with the scientists doing the work. He describes the dedication and expertise required to drill ice cores in remote locations and transport those intact cores to permanent storage and lab locations. He also provides a coherent and concise summary of the history of the climate change concept that I revisit in chapter 10.

But the most complete description of the process, excitement, and science of actually collecting and reading the stories in ice cores is *The Two-Mile Time Machine*, by Richard Alley. Alley spent summers beginning as a graduate student helping to retrieve and conduct field analyses of a two-mile-long ice core collected from atop the Greenland ice sheet.

Although the extent of ice sheets has waxed and waned during successive ice ages, there is some very old ice around, buried deep in the massive ice sheets in Greenland and Antarctica. So how do you collect an ice core, and how do you read the stories it can tell? Alley describes in detail the level of international cooperation needed and the very challenging logistical requirements.

The simple concept is to find a spot with the deepest and, one hopes, the oldest accumulation of ice. Greenland is ideal for this as snowfall is abundant in most years, so the amount of ice to be sampled per year or decade is greatest. But it also requires the longest core to extract the entire record.

Not only is snowfall abundant year-round (in most years), the physical composition of new snow changes from summer to winter. In summer, there is enough moisture in the air for ice to form on the surface at night (hoarfrost or frozen dewfall). The ice crystals created differ in structure from new snow, and since they do not form in winter, the accumulation of these crystals, visually detectable, create annual layers, much like the annual rings of wood growth in a tree.

As snow accumulates over the years, it begins to compress lower layers until snow becomes ice. As snow, the material is permeable by the gases in the atmosphere, and the composition of air within the snow is changeable. Once the snow turns to ice, bubbles of air are shut off from the atmosphere and remain frozen in place, creating a fixed record of the chemistry of the atmosphere when the ice was formed.

To sample this historical record, special drilling equipment was required and developed to extract cores in manageable increments. Cores needed to be removed without compaction or damage, without melting from the friction of the drill bit, and without damaging the hole or allowing it to collapse from the pressure exerted on the empty column by the surrounding ice, so that the drill could be lowered again for the next core. Alley describes all the technology, ingenuity, creativity, and just plain hard work that allowed the task to be completed.

Several different sites and different cores have been collected from the Greenland ice sheet. Alley worked as part of the GISP2 (Greenland Ice Sheet Project 2) team, a primarily American effort paired with a primarily European effort nearby. Having two replicate sites and cores allowed cross-checking for accuracy.

All in all it took five years (1988–1993) of working in the twelve-week window available each summer to drill through those two miles of ice between the surface and the bedrock at the GISP2 site. A collected core section could be analyzed immediately for certain characteristics. Cores were placed on a light table and dated according to visible layers of summer and winter ice. A quick method for measuring chemical composition by the ability of the ice to transmit an electrical charge (conductivity) identified eras of changing environmental conditions due to major disruptions downwind, like fires and volcanoes. As one example, detection of chemicals deposited in Greenland as the result of a major volcanic eruption in Iceland in 1783 resulted in a clear chemical marker and a check on visual dating techniques. Eruptions from earlier periods could also be seen.

One thing that both Bowen and Alley make clear is the dedication and energy of large numbers of scientists, technicians, field hands, and lab workers who derive satisfaction and some measure of joy from the incredibly tedious work carried out under very challenging circumstances. Alley in particular captures the culture of the remote field station where teamwork, ingenuity, and cooperation are key.

After measuring what could be seen and sampled directly while sitting in a deep snow pit at subzero temperatures for long hours, the core needed to be packed for transport to one or more labs in different countries around the world—again ensuring that the core is neither damaged nor allowed to come close to melting. At those labs, cores again need to be stored without compromising the integrity of the ice.

The full two-mile GISP2 core is an atmospheric time machine (hence the title of Alley's book, *The Two-Mile Time Machine*). With extreme care, the bubbles of gas trapped when the snow first turned to ice can be extracted and analyzed. It's like bringing one-hundred-thousand-year-old air into today's lab. And while the GISP2 ice core only goes back about a hundred and ten thousand years, capturing one turn of the Milankovitch cycle, subsequent cores from Antarctica and elsewhere (like the Vostok core used to create figure 6.2) can now take this atmospheric history back up to eight hundred thousand years.

For many years my office window looked out over one of those facilities used to store these ice cores—a huge freezer sitting on a parking pad next to the building. It was always intriguing to think how much of the Earth's environmental and climate history resided in that huge white box.

Measuring carbon dioxide in those bubbles once they are very carefully liberated from the frozen core is a straightforward process using standard methods. But how do you measure what the temperature was at the time that bubble was trapped?

The best and most consistent measure of temperature from ice cores, as well as from ocean and lake sediments, is using isotopes of oxygen.

The most abundant form of oxygen has eight protons, eight neutrons, and eight electrons (atomic number: eight, atomic weight, the sum of protons and neutrons, about sixteen). A small fraction of oxygen atoms have two extra neutrons (atomic number still eight, but atomic weight about eighteen). The different forms of the same element are called isotopes, and the differences in weight (or mass if you like) cause them to enter into physical or chemical reactions at different rates. Just how different depends on the temperature. So oxygen isotope ratios (sometimes you will see this expressed as the ratio of O^{18}/O^{16} or just delta O^{18}) can record the temperature at the time of the last reaction (freezing, photosynthesis, evaporation, whatever). This technique provides the temperature data in figure 6.2.

So evidence from the landscape and the oceans tells us that there have been recurring periods of advance and retreat of massive amounts of glacial ice. Calculations based on subtle changes in the Earth's orbit and orientation toward the sun predict a major cycle of about a hundred thousand years for the coming and going of the ice. Ice cores provide evidence of just how much the temperature changed, and how the chemistry of the atmosphere changed as well.

This understanding provides context for current rates of change in both temperature and carbon dioxide concentration, to which I return in chapter 10. Is there also a story in the ice that gives a hint as to how rapidly climate has changed in the past, and might change in the near future?

Yes. Combining this oxygen thermometer with other measurements and increasingly precise methods of isolating and dating ice core segments yield some possibly disturbing results. Temperature changes of as much as 8°C occurred in Greenland near the end of the last major glaciation, ending about twelve thousand years ago.

The results have been summarized by Alley and others as saying that the past two thousand years, in which most of the development of human culture has occurred, has been a period of unusually calm climate.

You might have already identified that the Greenland ice samples were collected at the northern end of the Atlantic meridional overturning circulation, or AMOC, and so might record more extreme shifts than those experienced globally. Globally averaged temperatures show much smaller and slower changes, at least until recently (figure 6.3).

But even minor shifts within this globally calm period have affected the human enterprise. Brian Fagan in *The Little Ice Age* describes the impacts of the relatively small decrease in global temperatures between the years 1400 and 1800. He links major shifts in agriculture, human populations, pandemics, and all sorts of other catastrophes to this latter cold period. In *Guns, Germs, and Steel*, Jared Diamond chronicles the settlement of Greenland (and probably North America) by the Norse culture during the warmer period from 900 to 1300, and the abandonment of those settlements as the climate cooled.

Our transition from weather to climate has shifted the focus from short-term atmospheric dynamics that control the weather to the longer-term forcing functions or boundary conditions that control the longer-term changes we call climate. The ice core record provides an invaluable window into this part of Earth's climate history. For the ice ages covering about the past eight hundred thousand years, orbital forcing has set the pace for the timing of glacial advance and retreat

Global Average Temperature Change

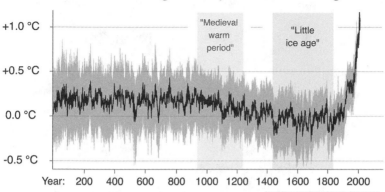

Figure 6.3. Changes in global average temperature over the past two thousand years.

globally, but feedbacks caused by changes in albedo and carbon dioxide have accentuated the temperature swings.

Ice cores and other surrogates for measuring climate offer the perspective that all of human culture has arisen at the top of the glacial cycle, during a period of extremely calm climate. Rapid shifts in the climate system appear to be more the rule than the exception, and they should perhaps be part of our thinking about current rates of climate change.

If we look at even longer time frames, we can ask this question: Just how far can these climate changes swing?

7

Extremes We Will Never See

The Whole Earth or Blue Marble images gifted to us by the Apollo astronauts on their way to the moon completely altered popular perceptions of the place of the Earth in the universe, and our relationship to the environment. If Rachel Carson's *Silent Spring* was the literary launch for the environmental movement of the 1960s and 1970s, these images were the icons.

In these images, the Earth looks like a rare and delicate blue ball against the dark and empty void of space (figure 7.1, top). Driven by solar energy, but formed by life, everything we experience in terms of climate and weather, and everything else, happens within an atmospheric envelope so thin as to be indistinguishable in either the Whole Earth images or closer views like that from the now defunct space shuttle (figure 7.1, bottom). A number of books written by astronauts combine these images with prose expressing awe and wonder at the sight. Often these explorers were left with a sense of the unique and fragile nature of our home planet.

Just how rare and precious is our blue ball, and how fragile? We can look out from the Earth's surface to the wider universe to answer that first question (rarity) and at both near-neighbor planets and our global climate history to address the second (fragility).

The goal of this brief excursion into deep time and space is to provide context for how our climate system is changing now, and perhaps to develop a sense of humility by knowing how very different

Figure 7.1. Two views of Earth from space; (top) a view of the Whole Earth or Blue Marble images captured by the Apollo astronauts on their way to the moon; (bottom) a closer view from the space shuttle.

our Earth might have been, and actually how different it was in the past. Our current climatic environment is neither preordained nor necessarily stable, nor is it the norm over geologic time.

We have actually located and identified a fair number of Earthlike planets. NASA's Exoplanet Exploration website updates the number daily, but at this writing there are more than four thousand in more than three thousand planetary systems (like our solar system). Discovery of these tiny, remote rocks is a wonder of modern technology. Most detections are based either on minute changes in solar irradiance as the planets cross in front of their suns (transit detection) or on even subtler changes in the kinds of light emitted by the host star as it shifts due to gravitational pulls from the orbiting planet (radial velocity detection).

But if you want to talk probabilities and big numbers, there are about two hundred billion stars just in our Milky Way galaxy, and something like 7 percent of them are thought to be of the same type as our sun. If each is calculated to have a 50 percent chance of an Earthlike planet in its solar system (perhaps a conservative estimate given four thousand identified in three thousand systems), there are several billion chances for life just in our own galaxy.

For these rocks to support life, current theory holds that they must be in the Goldilocks zone—not too close to and not too far from their sun, such that the climate would be moderate enough that liquid water could exist. Applying James Lovelock's definition of life in chapter 2, we cannot yet determine if any of these planets exhibit atmospheres maintained in chemical disequilibrium by life. Maybe someday we will.

Even if we were to discover chemical disequilibrium and life on other planets, looking for intelligent life takes this search to another level. By current theory, that would involve searching the skies for radio waves that have patterns suggesting purposeful creation. The

SETI (Search for Extraterrestrial Intelligence) program pursued this concept for many years but was essentially shut down in 2020 when the largest installation for capturing radio waves, the Arecibo radio telescope in Puerto Rico, collapsed. The 1997 movie *Contact* captures some of this effort with a sci-fi conclusion. If you are looking for thoughts on the durability or persistence of intelligent life, you can search for the Drake Equation, or read the last two chapters in the original version of Carl Sagan's *Cosmos*.

To think about the potential for life on those four thousand identified or billions of possible planets, we can look to our own nearest neighbors, both of them in the Goldilocks zone, for clues how planetary environments might change in ways that would make them look very different from Earth.

In *Caesar's Last Breath*, Sam Kean describes plainly how Earth, Venus, and Mars all had the potential to harbor Earthlike atmospheres, with volcanic activity creating the basic gases required and sufficient gravity retaining some of those gases against the explosive impacts with passing comets, asteroids, and planetoids that characterized the early eons of our solar system. Mars failed to keep its atmosphere and, without that protective layer, became barren and cold, showing huge daily swings in temperature. Venus, on the other hand, being closer to the sun was too hot to allow water vapor spewed by volcanoes to condense, so there was no liquid water to host the reactions that would absorb carbon dioxide. The result is crushing atmospheric pressures at the surface equal to roughly a kilometer deep in the oceans on Earth, a concentration of carbon gases two hundred times that on Earth, and a temperature above 450°C.

So it may take more than just being a Goldilocks planet to actually have liquid water, and life. But being dead now does not mean this was always the case. At this writing, three different space missions developed by China, the United Arab Emirates, and the United States are approaching Mars to see if perhaps there was water on the surface and life had begun there early in its history. And in 2020 some claimed to detect phosphine gas in the atmosphere of Venus, which reportedly should not occur in the absence of life.

Has Earth ever approached these extremes? I will leave it to you to look elsewhere for the answers about the very early Earth. It was certainly Hades hot (and the geologic period is called Hadean) over the first several hundred million years, especially after a Mars-sized object crashed into the Earth and ripped out a huge piece that became our moon (at least as current theory holds). As planets have swept their neighborhoods clear of other objects, adding to their own mass, such collisions have become less common. Through these upheavals, Earth's atmosphere has passed through four stages with very different chemistries. Kean (and others) have described these, and the first three are too different from our current environment to provide insights into our climate system, so we will confine ourselves to the past billion years or so.

With Lovelock's definition of life as a process that uses solar energy to create an atmosphere in disequilibrium, ours is the last of the four atmospheres, composed almost entirely of diatomic forms of nitrogen and oxygen (N_2 and O_2), that would signal to any distant observers the presence of life. On Earth, the biology of photosynthesis is what sustains that high concentration of oxygen.

Lovelock's first edition of that little gem *Gaia* not only gives us his unique definition of life at the planetary scale (disequilibrium) that made him unpopular at NASA. While writing at the height of the first round of environmental awareness in the 1960s, he put biodiversity and pollution issues in perspective by describing the greatest environmental catastrophe and mass extinction event to hit life on Earth—the creation of that oxygen-rich atmosphere.

It took nearly four billion years for chemical and physical processes and the evolution of effective means of harvesting solar energy for life through photosynthesis to wreak this havoc on the world. Not only was photosynthesis required, the organic matter created through photosynthesis also needed to be stored away in a place where decomposition could not use that oxygen to reverse the process. For example, organic matter drifting to the bottom of the oceans or deposited in swamps and bogs could remain undecomposed for eons, leaving the waste product of photosynthesis, free oxygen, in the atmosphere.

While small amounts of oxygen were created in earlier eras by anaerobic bacteria carrying out photosynthesis, it was not until about six hundred million years ago that significant amounts of oxygen accumulated (figure 7.2). It is no coincidence that what is termed the Cambrian explosion, signaling the evolution of multicellular life, including higher plants and animals, and eventually us, followed Lovelock's greatest environmental disaster, when dominance of life forms switched from those that do not require oxygen to those that do (again, like us).

Oxygen is a highly reactive gas, which is why it is so useful in the energetics of higher animals (again, like us), and would not be present in the atmosphere at the level of 21 percent that it is now were it not being continually produced by photosynthesis. Another Lovelock example worth reading in *Gaia* is a description of feedback controls that tend to maintain oxygen concentrations at that level. For example, at not much above 21 percent, spontaneous combustion of organic matter (like humus in soils, and wood in trees) would quickly bring any excursion in the concentration of this gas back down to current levels.

A word of warning: if you start with a Web search on Gaia, you will be treated to a wealth of pseudoscience, mythology, and other fanciful pursuits. Gaia is not an organism, and Lovelock initially did not intend it to be much more than a metaphor, and a useful one, for describing the kinds of negative feedbacks that might have been important in stabilizing Earth's environment and allowing life to persist for billions of years.

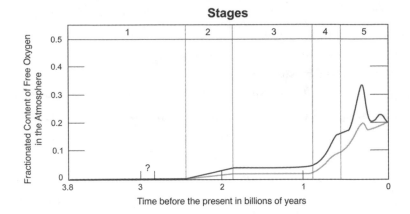

Figure 7.2. The concentration of oxygen in the Earth's atmosphere over geologic time. On the concentration axis, 0.21 represents the current concentration of 21 percent. The two lines represent estimates from two different studies.

Since biology has determined the state of the atmosphere since the Cambrian explosion, how much has global climate changed over the past six hundred million years?

I like the way figure 7.3 tells this story. Both the time scale and the degree of detail change as we look back, first tens of thousands to hundreds of thousands of years on the right, and then millions and hundreds of millions of years on the left.

How do we know what the temperature was three hundred million years ago? The most frequently used paleothermometer is the same ratio of isotopes of oxygen applied to ice cores as were captured by marine organisms and preserved in ocean sediments. Additional evidence comes from the distribution of fossils of organisms adapted to different temperature regimes.

A few points we can take from figure 7.3 are that ancient changes in temperature have been extreme (shown in the far left panel) and that there has been a general cooling trend over the past fifty million

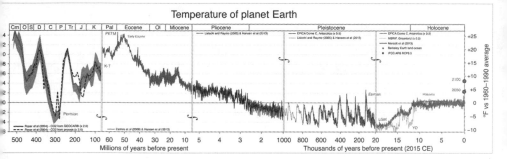

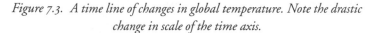

Figure 7.3. A time line of changes in global temperature. Note the drastic change in scale of the time axis.

years. With the finer-scale measurements possible from ice cores, the variations within this trend due to the Milankovitch cycle (the ice ages) can be seen on the right. Even with this variable resolution, you can see that our recent past represents an unusually moderate and stable period.

We are getting ahead of ourselves a bit here, but those little dots on the far right axis show two predictions for changes in global temperature by 2050 and 2100. The higher dot would represent a global temperature higher than any in the past ten million years.

Causes for the very large swings in temperature are linked primarily to changes in the concentrations of greenhouse gases, especially carbon dioxide and methane. The former may have reached concentrations as much as two to five times the current concentration of just over four hundred parts per million. Shifts in landmasses through plate tectonics can also affect the absorption of sunlight and the existence of ocean currents that redistribute that energy, as well as landmasses capable of supporting huge ice caps.

Carbon dioxide increases are linked to higher volcanic activity, but the higher temperatures, in a positive feedback, also lead to warmer oceans, and warmer water absorbs less carbon dioxide from the atmosphere. Warmer temperatures can also lead to increased methane release.

Temperature changes captured in figure 7.3 seem extreme to us in terms of the climate that we now have. And remember again that even the small changes of less than 1°C in the current interglacial era have impacted the human experience significantly.

What this does say is that, at least since life came to dominate the atmosphere, the Earth has never approached the superheated state that Venus occupies, and it seems very unlikely to ever go that way.

What about the cold end of the range? The ice ages show large and rapid shifts in average temperature. Might we go colder? This was a concern that Arrhenius held, living in the far north and in a scientific milieu that had only recently accepted the existence of ice ages. His comments about "a more equable climate" with increasing carbon dioxide concentrations through industrial activity reflect that time and place. Even more recently, a class I took as a graduate student in the 1970s from a very reputable scientist spent as much time on possible ice ages as on global warming. This was a reflection of the times, in that global temperatures had then plateaued for a short period. That time seems very far away now.

There has been a lot of press over the past several years (well, science press anyway) given to the concept of Snowball Earth—a time before the Cambrian explosion when the Earth seems to have come perilously close to going permanently ice covered. As it is a recent topic related to an extremely distant event, there are still several lines of evidence that dispute this ever happened, and many mechanisms proposed for why it did. The Wikipedia page on the topic should be under active revision for many years as the scientific communities engaged in the discussion have their say.

Evidence in favor of Snowball Earth includes primarily the existence of glacial deposits of the right age found in places that would have been equatorial at the time of formation. This assumes an understanding of plate tectonics, paleoclimates, and the existence, hundreds of millions of years after deposition, of readable geologic formations. One source charts not one but up to three Snowball Earth events prior to the Cambrian explosion.

Rather than go deeper into the evidence for or against, let's accept that this occurred. What would drive the Earth to such extremes, and what could possibly bring it back?

The reason to include this event, and this entire chapter, is to emphasize the role of feedbacks at the planetary level in driving changes in climate. One line of reasoning suggests that shifting continents due to plate tectonics led to uplifting of mountains high enough to allow glacial formation, and also shifts in ocean currents that would restrict the redistribution of tropical heat toward the poles. Concentrations of greenhouse gases were low, limiting their ability to warm the atmosphere. As ice began to form, the positive feedback described earlier involving increased reflection of light (albedo) kicked in, leading to further cooling and the formation of more ice.

All of this seems very reasonable, but once complete or nearly complete ice cover occurs (and some describe more of a Slushball Earth), what could bring the planet back to a warmer state? A Russian climatologist named Mikhail Budyko produced a simple model of global climate driven by initial ice formation and the feedbacks caused by changes in albedo. He suggested that the Earth should be permanently in either a completely ice-covered or a totally ice-free condition; that anything in between was unstable. Based on this model, Budyko concluded, according to one source, that Snowball Earth should never have happened, as he could see no way for a return to a reduced ice condition.

One straightforward mechanism that would explain the shift back to warmer times is that volcanism continued to occur, with gases like carbon dioxide and water vapor ejected above the ice and into the atmosphere. With ice covering both land and sea, carbon dioxide could not be absorbed into seawater or be removed by reactions with rocks. As the greenhouse gases accumulated, warming would occur, and the ice would eventually retreat. As the ice retreated, the positive feedback would work in the opposite direction, with open water absorbing more sunlight (low albedo), leading to a warmer world.

The real point to be made here relative to our major focus on more recent weather and climate controls is that positive and negative feedbacks are important and powerful. Combining this fact with the realization that climatic changes large enough to affect the way we live have occurred in historic times, and that larger changes are predicted in the future (getting ahead of ourselves again), might affect how we think about and approach questions related to climate. Perhaps our mind-set should not be to assume that climates are inherently stable (they aren't) or that changes we might initiate won't be multiplied through feedback effects (they might).

A healthy respect for the ability of climate regimes to change quickly might be warranted, as well as a bit of humility in how we think about our climate system.

With this foundation on the workings of the climate/weather energy machine at very short to very long time frames, we can move on to room 3 of our tour, shifting the focus to how Earth system science gets done, and how that approach to science has helped solve some important environmental issues.

ROOM THREE

Earth System Science as a Process

HOW IT WORKS, AND WORKS FOR US

8

How Long Does It Take to Shift a Paradigm?

History and the history of science tend to use a shorthand that equates a particular breakthrough with a particular person. Charles Darwin and evolution, Albert Einstein and relativity, Marie Curie and radioactivity, and (I have been somewhat guilty of this myself) Arrhenius and carbon dioxide and climate. But I hope our tour of the Stockholm Physics Society serves as one example of the fact that major breakthroughs do not occur in an intellectual vacuum. Every detailed story of a shifting paradigm, or major advance in scientific thinking, will demonstrate that the genius with whom we match a major advance lived and worked in a time, and with a community of thinkers, that supported and enabled that big advance.

But how does that shift in worldview happen? In a recent and insightful analysis of the origins and power of the scientific method, *The Knowledge Machine*, Michael Strevens constructs a framework that describes that power. Its primary source is what he terms the Iron Rule of Explanation, the idea that all arguments are to be settled by empirical testing. Corollaries include a separation of scientific reasoning from other lines of explanation (religion, philosophy, aesthetics), a striving for objectivity, and a limiting of public scientific presentations to matters of fact only. Speculation and creativity are limited to private discussions or scientific meetings.

Strevens also sets up a dichotomy that addresses our question here on shifting paradigms. He compares the view associated with Karl

Popper that science progresses continually by establishing hypotheses that can be disproved by empirical testing, with the notion of scientific revolutions associated with Thomas Kuhn (see the caveats above about "big name" shorthand!).

Essentially, Popper's view has been termed "normal science," where researchers working within a certain worldview make incremental improvements in understanding by using that worldview to frame questions and hypotheses. One tenet of this concept is that hypotheses can never be absolutely proven but alternatives can be disproven by observation and measurement; something like the sentiment ascribed to Sherlock Holmes by Arthur Conan Doyle: once you eliminate the impossible, whatever remains, no matter how improbable, must be the truth.

Thomas Kuhn's influential work *The Structure of Scientific Revolutions* describes those revolutions as paradigm shifts occurring in a set of stages. When normal science begins to accumulate significant contradictions to the existing worldview, the field is thrown into a chaotic state where new ideas and "extraordinary research" can come into play as efforts are made to resolve the contradictions. Through this chaotic period, a new paradigm can develop that, in one integrated story, explains all (or most) of what appeared to be contradictory evidence.

It is in this chaotic phase that Strevens's Iron Rule is so important. Nothing can be dismissed just because it contradicts the current worldview. An idea, concept, or hypothesis can only be dismissed if it contradicts repeated, verifiable measurement.

We will see in the following examples that extraordinary research (meaning work outside the existing framework or worldview) can originate in a wide range of disciplines beyond those that might have been considered normal for that idea, and that economics and technology can play important roles. For Earth system science, our ability to see the world in new ways is a central force driving increased understanding.

For example, coal mines and canals created in the early years of the industrial revolution revealed geological strata allowing a generalized view of Earth's history (for a great telling of this story, and the role of a humble and unconventional engineer in the process, read *The Map That Changed the World*, by Simon Winchester). The discovery of radioactivity and radioactive decay led to power stations and bombs but also to better ways to estimate the age of the Earth. Understanding the nature of photons and light has many commercial applications but also allows calculations of the energy balance of the Earth and the monitoring of the Earth system by remote sensing satellites. The list of examples is endless, really. Science, technology, economic development, and, sadly, warfare, go hand in hand.

There is one observation Strevens makes that I find curious, and counter to my own experience. He characterizes normal research as drudgery. The kind of mind-numbing, repetitive, soul-killing activity that no one would undertake without the promise of success and wealth at the end. While he is right on so many other fronts, I would respectfully disagree on this one. What Strevens misses in this is the concept of flow in work—to love the process.

In psychology, flow has been described as arising from a love of the work itself and from being totally absorbed in the process. In this state, thought and action merge, leading to optimal performance, but also to a sense of a suspension of time. This captures what Richard Alley experienced working under incredibly arduous conditions in Greenland, and what Mark Bowen captures on mountaintop glaciers in the tropics. In a small way, I have experienced that myself, when a day analyzing data in spreadsheets—talk about boring drudgery— was absorbing because of the fun of the work process, the challenge of solving the puzzle presented by the data in hand, and an abiding intellectual interest in the eventual outcome.

It is hard to see that the monumental calculations we have attributed to Arrhenius, Walker, and Milankovitch, among others, were seen as drudgery for personal gain. Being intrigued by the problem and driven by the need to solve it seems the more powerful motivation. When advising potential Ph.D. students, I have been known to say that they should only go through that intense and tightly focused process if they can't help themselves; if this is what they always think about anyway.

Strevens does present one analogy of the scientific process that I find particularly compelling. He likens scientific progress to the construction of a coral reef. Millions of individual polyps on the living reef pursue the messy processes of life and death, but out of this process emerges the stable, permanent structure of the reef itself. Scientists engage in the dynamic and very human process of give-and-take, measurement and presentation, discussion and disputation, but, as Strevens says, what emerges is a solid and largely permanent structure of basic understanding.

So how does this relate to shifting paradigms for some of the big questions in Earth system science? Let's explore three: How old is the Earth? Do the continents move? Did a meteor strike kill off the dinosaurs? Although many other big questions could be included, these three show the emergent pattern that we are after. For each, the emphasis is on the steps and time lines for discovery, controversy, and eventual resolution, and how they compare. I repeat this approach in chapter 9, in discussing the stages of environmental grief as related to the discovery, controversy, and resolution of some major environmental issues.

The brief summaries offered here focus on the evidence accumulated over time and the different approaches taken, not on the interactions with the social or religious milieu in which they occurred. This will be a What do we know? and How do we know it? approach. Tomes have been written on the conflicts between science and religion, and any good biography of either Charles Darwin or Galileo Galilei will give prime examples (I especially like Dava Sobel's *Galileo's Daughter* for a unique view of his life and times).

Our focus, then, is neither the harsh give-and-take and criticism that is at the heart of the scientific process nor the social context of the arguments but rather the ideas and measurements brought to the topic from a wide diversity of fields of study, and their eventual coalescence into a new, and possibly final, view of one part of the Earth system. To use Strevens's analogy, I will emphasize the reef, while focusing on only a few of the polyps. As we will see, the time line for this process varies widely from one concept to another.

*

Creation myths are common to most cultures, and the Judeo-Christian story of the seven days in which the world was made is but one. Our telling here of the time line for the discovery of the age of the Earth begins with this creation story only because it provides the basis for the first estimation of the age of the Earth in the Western tradition. Many other traditions present far more complex stories of the creation of the world, with longer or undefinable time lines than the simplified, numerical calculation I describe here.

This Western story begins with a seventeenth-century scholar and theologian, James Ussher, Anglican archbishop of Armagh in Ireland. What sources did he have for his calculation? In the absence of the wealth of scientific observations we now take for granted, he used histories and genealogies in the Old Testament and other religious tracts to develop specific chronologies for a number of events, both historical and mythical. From these sources he calculated a precise time and date for creation of Earth: nightfall on October 22, 4004 B.C. That date can still be encountered in the modern world.

Although this estimate is now dismissed as simplistic as well as erroneous, no less an authority than Stephen J. Gould has defended Ussher's chronology as an honorable effort, saying that his work represented the best of scholarship in his time as part of a community of intellectuals working toward a common goal using accepted methods.

And there is another characteristic here that we have seen in the work of Milankovitch and Arrhenius—painstaking, arduous, manual computations based on available data.

The concept of a six-thousand-year old Earth held sway in European thought until new observations and technologies presented unresolvable conflicts with that time line. Simple but insightful observations, such as those of Nicolas Steno in the mid-1600s, led to questions about how sharks' teeth were to be found on mountaintops and embedded within hard rocks. He is credited with initiating the concepts of stratigraphy and the sequential formation of sediments, and beginning to question the universal explanation that the biblical flood (Noah and the Ark) caused all the visible forms of the Earth.

Direct observation of Earth's history became unavoidable with the initiation of massive projects to mine coal and build canals as part of the industrial revolution. All of this digging led to discoveries of the universal occurrence of geological strata across large swaths of the landscape (see again *The Map That Changed the World*), and to initial concepts of the formation of rocks by erosion and sedimentation. Such observations led James Hutton, familiar with both farming and the construction of canals, to consider a much older world.

Iconic in this oversimplified version of the story is the epiphanic moment in the recognition of deep time in what has come to be called Hutton's Unconformity. An unconformity in the layers or strata of exposed rocks occurs when one series of deposited sediments overlies another at a very different angle. With only general ideas of how long the formation of such strata would require, Hutton famously wrote,

"Here are three distinct successive periods of existence, and each of these is, in our measurement of time, a thing of infinite duration. . . . The result, therefore, of this physical inquiry is, that we find no vestige of a beginning, no prospect of an end."

The year was 1788, and the best guess on the age of the Earth went from about six thousand years to infinite!

Hutton became a central figure in the Scottish Enlightenment, and a colleague in that brilliant epoch, John Playfair, when exposed to this geological evidence, added to Hutton's conjecture: "The mind seemed to grow giddy by looking so far back into the abyss of time . . . we became sensible how much further reason may sometimes go than imagination may venture to follow."

But just how old, really? Charles Darwin applied his creative genius to a great many subjects beyond evolution. From earthworms to geology, he allowed his observations to lead to unusual conclusions. One of the earliest estimations of the rate of rock strata formation was based on his observations of the rate of sedimentation in a local estuary. Comparing current rates of erosion and sedimentation with the depth of sediments, he calculated that they had been accumulating for three hundred million years.

Critical to these calculations was the idea that the rates of sedimentation observed by Darwin had been the same over that entire three-hundred-million-year history. This concept, termed uniformitarianism, became associated with Charles Lyell, who produced twelve editions of the classic *Principles of Geology* between 1830 and 1875, one of the best-selling books of its time. Uniformitarianism was proposed in direct opposition to the previously prevailing view that all current features on the Earth were the result of original creation and the devastation of the biblical flood, a concept labeled catastrophism. Uniformitarianism argued that the world was very old, but without providing a way to judge just how old.

Despite this lack of precision, the concept was crucial for the young Darwin, who took an early version of Lyell's book with him on the famous voyage of HMS *Beagle* (1831–1836) that led him to the theory of evolution by natural selection. Darwin needed a nearly infinite time line to allow for the creation of the visible diversity of life forms he encountered.

Through the preeminence of Lyell and later geologists, uniformitarianism came to be the ruling paradigm in the discipline. It was now time for a genius from a very different field to weigh in.

William Thompson, first Baron Kelvin, was a physicist, engineer, and inventor of genius proportions. He is responsible for the Kelvin scale of temperature measurement, helped design power stations, and was crucial to laying the first transatlantic telegraphic cable. He was made a lord for his efforts. As a founder of the modern concepts of thermodynamics, he brought the rigor of physics to bear on this question of the age of the Earth.

His approach was based on the initial temperature of the Earth at its formation, the rate of conduction of heat energy toward the surface, and its radiation into space. It was the existence of deep mines, and the temperature gradient measured with depth, that allowed the estimate of the rate of heat loss to the surface. Using these measurements and the unassailable laws of physics, Kelvin calculated that the Earth could not be more than a hundred million years old, an estimate he later reduced to twenty to forty million years.

This is a classic confrontation in Earth system science. Two very different disciplines based on the best available evidence and the central concepts of their fields have come to totally different conclusions!

So how were these different estimates reconciled, and just how old is the Earth? Enter an entirely new set of researchers, and a previously unknown process.

Marie Skłodowska-Curie (Madam Curie) was the first woman to win a Nobel Prize, the first person to win two Nobels, and the only person to win Nobels in two different categories (Chemistry and Physics). She coined the term "radioactivity" and was a principal developer of the theory of radioactivity. She discovered two new elements, polonium and radium.

This incredible research productivity opened many new areas of science and technology, and, as with Darwin, the relevance to the question of the age of the Earth was an interesting sidelight. Later research has revealed that the radioactive decay of elements like uranium and thorium in the Earth's mantle and crust (the upper layers of the solid Earth) continuously generate forty-four trillion watts of heat. The rigorous calculations by Kelvin were not wrong, but they did not include a previously huge, unknown source of heat energy that alters the basic calculation.

The theory of radioactive decay also provides the best current estimate of the age of the Earth. Different radioactive elements decay at different rates. This decay converts the initial element (like radioactive uranium) into a stable element (like lead). The rate at which this occurs can be measured with exquisite precision. By knowing the initial composition of materials and the current relative concentration of both the radioactive source and the stable product of decay, the age of that material can be estimated.

Applying this concept to the question of the age of the Earth requires finding rocks that have not been modified since formation, a daunting task. Augmenting this with samples from comets and asteroids, which have remained unchanged since the formation of our solar system, the current age of the Earth is estimated at 4.5 billion years, give or take a few tens of millions of years.

Although that number is interesting and creates enough time for other initially controversial processes, like evolution, to happen, it is the give-and-take among different disciplines and experts that is our focus. Historical documents, emerging technologies, basic geology, thermodynamics, and nuclear physics all played a role in this story.

*

A child looks at a modern globe and notices some features—for example, how much water there is and the shapes of the continents—and, being genetically programmed for pattern recognition, also notices that two of the continents have complementary shapes. Isn't it interesting that Africa and South America look almost like pieces of a puzzle—as if they could be pushed back together into a single shape? Maybe you have done that yourself? I remember doing it as a child.

If you made that simple observation, you are not alone. Benjamin Franklin, Francis Bacon, and Alexander von Humboldt all expressed the same idea. In 1596, a Belgian mapmaker named Abraham Ortelius said that the Americas were "torn away from Europe and Africa . . . by earthquakes and floods" based on the complementary shapes of the three continents. This observation and conclusion arose nearly as soon as maps were available that were accurate enough to make the comparison.

But of course things as big as continents can't actually move, right? Ridiculous idea.

Yet in 1915 Alfred Wegener proposed just that. Recognizing those who had proposed this before him, in his book *The Origin of Continents and Oceans* he used not only the outline of the continents but also the dovetailing of geologic formations on either side of the Atlantic to support what he called continental drift. At about the same time, an American named Frank Taylor noted a similar convergence among fossil plant species on both sides (figure 8.1).

Like the geologists in the controversy about the age of the Earth, those proposing that continental drift happens had lots of circumstantial evidence in the distribution of rock types and fossils. And like

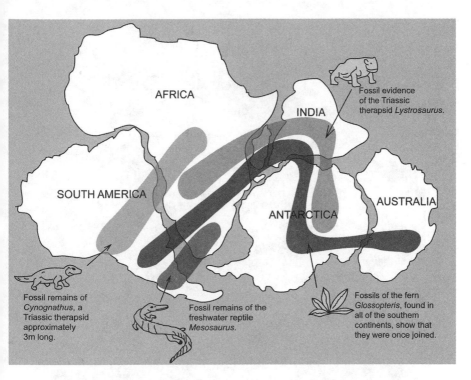

Figure 8.1. A reconstruction of the similarities among fossil species across existing continents used by Alfred Wegener to support the idea that the continents were once united.

those geologists, they lacked a physical process that could explain the monumental shifts in continents and provide a source of energy that could accomplish such a feat. These questions remained unanswerable until new ways of seeing the Earth arose after World War II.

In the 1950s, technological innovations began to allow detailed mapping of the ocean floor and revealed the largest geologic feature on the planet. The Mid-Atlantic Ridge (figure 8.2) is a twenty-four-thousand-mile-long underwater mountain chain running basically down the center of the Atlantic Ocean, from north of Iceland to near Antarctica. Separate ocean-crossing voyages had glimpsed this ridge,

Figure 8.2. A map of ocean depth showing the Mid-Atlantic Ridge, indicated by the lighter gray feature.

but it was not until sonar- and laser-based technologies allowed rapid coverage of large areas that the full extent and continuity of the ridge were established.

What could cause such a feature? There was evidence of earthquake-like activity along the ridge, suggesting movement, but definitive evidence came from a combination of radioactive dating and another new method, with the great party-stopping name of paleomagnetism.

North is always "up," right? Well, no. We take the position of the north and south poles for granted (although careful explorers know that the geographic north pole—the axis around which the Earth turns—is not the same as the magnetic north pole). But not only do the positions of the magnetic north and south poles wander, sometimes the poles switch places!

How can we know such a thing? Minerals called magnetites contain iron elements that orient with and record the direction of the magnetic fields, and which way is up, when they become fixed in cooling lava. Basically, lava flows from volcanoes are melted rock. When the lava cools, new rocks (basalts) are formed. As the material cools and solidifies, the magnetites orient toward the poles, providing a compass frozen in time.

Magnetites actually provide one example of confusing data that became resolved by a new paradigm. Magnetites were recognized and mapped before the drifting of continents was accepted. Magnetites of different ages were arrayed in very different directions, seeming to suggest that the location of the magnetic poles had not just wandered but had migrated thousands of kilometers over millions of years. There was no explanation for such shifts in the orientation of magnetic north and south.

When the technology of underwater exploration finally enabled the collection of rock samples from the seafloor up to 8 km below the surface, transects or lines of samples could be collected at different distances from the center of the Mid-Atlantic Ridge. Analyzing the samples for the orientation of magnetites documented the surprising fact that the magnetic poles had actually reversed locations, and not just once but many, many times.

Not only was "up" not always up, but our north became south, and vice versa, with some frequency (figure 8.3). Similar patterns had been found on surface rocks and been dated using radiometric techniques, so an age could be associated with each polar reversal.

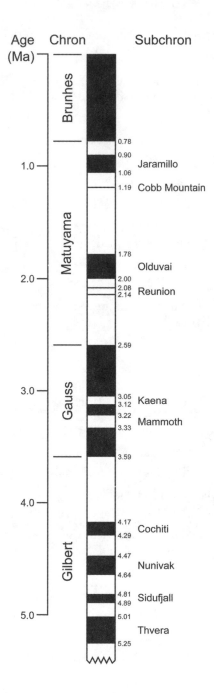

Figure 8.3. An example of changes in orientation of magnetites in response to the reversal of the north and south poles. Each switch from dark to light or back marks a switch in the orientation of the magnetic poles. Age (Ma) is in millions of years.

A clear and intriguing data set on the dating of reversals and rates of spread along the Mid-Atlantic Ridge made for another simple spreadsheet lab in my class, this time showing a consistent rate of spread both east and west. Europe and North America are drifting apart by about 2 cm per year. Dating the timing of the polar reversals also allows a calculation of how frequently they occur, which appears, on average, to be once every two hundred thousand to three hundred thousand years. If you want another party-stopping piece of jargon, you can call on the technical name for this branch of science: reversal paleomagnetostratigraphy.

This proposed paradigm shift, that continents move at measurable rates, had the usual consequences. It opened the door to completely new sets of observations and research that led to a detailed understanding, and even mapping, of the movement of continents (excellent videos of the history of these movements going back more than a billion years are available online).

How does this shift resolve previously confounding observations? The list is too long to include here, but, for example, major tree genera (plural of genus) such as oak and maple are shared across China, North America, and Europe. Fossil evidence shows that they evolved about a hundred and fifty million years ago, when the three continents were still joined. As the continents moved apart, the genera of oak, maple, and other trees retained the major characteristics but evolved into the different species we now see. If you are planting a Japanese maple in your front yard (or see one in a park), think that this now-foreign tree species shares a genetic origin with the native sugar maples you might see in the woods or in that same park.

Similar stories can be told for geological formations, fossils, and other recorders of Earth's history. For example, the Scottish Highlands and the Appalachian Mountains of the eastern United States share a common origin (about four hundred million years ago) but are now separated as Europe and North America move away from each other.

And finally, the orientation of magnetites that initially suggested extreme polar wandering were brought into alignment with the actual location of the poles by understanding that it was not the magnetic poles that had wandered widely but rather the continents from which those rock samples had been collected.

And where is the energy for this movement coming from, and how does the movement happen? Will you believe that it is the same concept that leads to thunderstorms or causes water to boil up from the bottom of a pan when heated? Yes, the process is convection.

It is hard to think of the solid Earth as not entirely solid, but the surface on which we tread (called the lithosphere), which includes both continental and oceanic crusts, is underlain by a superheated, only semisolid layer called the asthenosphere (another great party word) that cycles in response to heating from below—transferring heat from lower layers by convection. The same process that allows us to date the age of the Earth (radioactive decay) and dismissed Kelvin's pure science calculations also provides the heat energy that drives plate tectonics (figure 8.4).

So let's count the traditional disciplines that led us to plate tectonics. Observational biology (distribution of species) and geology (distribution of rock types) suggested to Wegner, Taylor, and others that the continents were once linked. Physical oceanography identified a major geological feature (the Mid-Atlantic Ridge) that led to research on paleomagnetism and dating of sediments by radioactive element decay. Many others could be described, but the point here is that Earth system science is best pursued as a seamless integration of all of the traditional biological, chemical, and physical sciences—kind of like what happened in the Stockholm Physics Society.

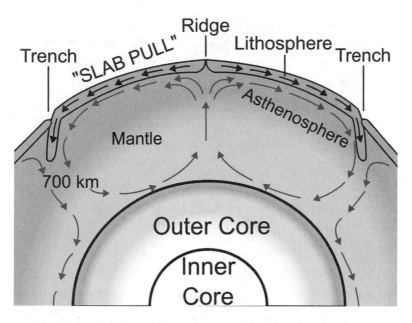

Figure 8.4. A depiction of the movement of the lithosphere (oceanic and continental crusts) "floating" on the asthenosphere, a region of semisolid material driven in convective cycles by heat generated in the Earth's core.

And final resolution of the issue is more recent than one might think for such a now-basic truth about our home planet. As a graduate student first exposed to the concept in the mid-1970s, I assumed, since it was presented to me as scientific gospel, that this had been known forever. I was disabused of that notion at a seminar a few years later in which the time line was presented, and the fact that no more than ten years earlier the concept had still been in dispute.

*

From Ussher through Lyell and into the twentieth century, a central controversy about the processes forming the Earth was over the warring concepts of uniformitarianism versus catastrophism—or the question of whether what we see on the Earth today reflects the long accumulation of processes that are currently in action (erosion, sedimentation, plate tectonics) or have resulted from sudden, major catastrophic events.

To support the notion of an Earth that is six thousand years old, one needed the biblical flood (Noah and the Ark) to explain phenomena such as marine fossils on mountaintops. Overcoming the social and religious inertia supporting this worldview was a monumental paradigm shift, hard won, which Lyell, Darwin, and others supported, and which was required to explain their observations. Uniformitarianism was the rule of day well into the twentieth century.

And then there was the discovery of the K/T meteor and the extinction of the dinosaurs.

Dinosaurs—imagine digging at an archaeological site or for geological research and coming across a bone many, many times larger than any present-day animal could support. The history of the discovery of such bones in the 1830s, followed by the Bone Wars waged by different groups of entrepreneurial diggers in pursuit of fame and fortune more than science, makes fascinating reading. For decades, really more than a century, both scientists and entrepreneurs were engaged in discovering, piecing together, and trying to understand a world that could support such massive creatures.

Clearly there is nothing like them in the world today. What happened to them? Oddly enough, the question did not rise to the top of the list for more than a century.

Which brings us to one of the best scientific mystery stories, in my opinion, ever told.

In *T. rex and the Crater of Doom*, Walter Alvarez tells the compelling story of the full process of determining just what was responsible for a global mass extinction event sixty-five million years ago, including the disappearance of the dinosaurs. It reads like a good mystery novel, with major clues, dead-end searches, and a final resolution. In this case, the criminal caught in the act was not a thief but the concept of uniformitarianism, and the weapon was a massive meteor.

The full story is as clear a presentation of the steps in a paradigm shift as you can encounter. Alvarez gives it a rich telling that can only be highlighted here. My focus is on the parts of the story that relate to our goal in this chapter: demonstrating the role of creative approaches and expertise from many different areas of science in solving a major scientific riddle.

The story begins with two young, recently minted Ph.D. geologists, Alvarez and a colleague, Bill Lowrie, trained in classical geology, imbued with the concept of uniformitarianism and aware of the recent discoveries in magnetic reversals, collecting rock samples in the Italian Apennines. Their initial goal, in the mid-1970s, was to add this location to the growing set of geologic strata revealing the timing of the switches in the magnetic poles.

Geologic ages are separated as much by changes in the fossil record of species present as by changes in the rocks themselves. The fact that biology has, in effect, ruled the definition of the major geologic eras has always been fascinating to me.

As well-trained geologists, Alvarez and Lowrie were aware of the importance of fossils, and as they worked down their strata and back through time, they encountered the boundary that marks the end of the Cretaceous age and the beginning of the Tertiary, dated to about sixty-five million years ago. That boundary was marked, in part, by a nearly complete turnover in the species of marine organisms known as Foraminifera, or forams.

This was the mass extinction, as we now call these events, that marked the boundary between these two geologic eras. At the rock face Alvarez and Lowrie were sampling, the formations of limestone above and below this epic change were similar, supporting the uniformitarian view of continuity of geologic processes before and after.

Precisely at this boundary, however, Alvarez noted the existence of a dense layer of clay, totally distinct from the rocks above and below.

What had happened in between those two limestone layers that caused the mass extinction of forams, and was that somehow related to this odd clay layer? When Alvarez attended a seminar where the speaker noted that the most famous mass extinction of all time—the disappearance of the dinosaurs—happened at about the same time, he was hooked.

Alvarez recounts, with some amazement, that mass extinctions, including the disappearance of the dinosaurs, were a nonproblem during his time as a student. Adhering to uniformitarian principles, the extinction was thought to have occurred over millions of years and be due to climate change, altered sea levels, and the like. Alvarez notes that, into the 1980s, the few suggestions that a catastrophic event might have eliminated the largest land animals ever to walk the planet were treated dismissively, and the issue remained unimportant.

What follows is a truly superficial treatment of the rich and dynamic interactions that led to the actual source of that layer of clay, and the disappearance of the dinosaurs, but it is the best that can be done here. I strongly recommend the book!

Alvarez remained intrigued by that K/T clay layer (the K is from the German spelling of Cretaceous) and wanted to know more. He collected samples of the clay and wondered what kind of information could be gained from them. He reports feeling that the answers would come from outside the field of classical geology.

Alvarez's father, Lewis Alvarez, was an accomplished physicist who had helped identify a host of subatomic particles and was familiar with recent advances in measurement of infinitesimally low concentrations of rare elements through a technology called neutron activation. Walter Alvarez wanted to know how long it had taken for this K/T clay layer to form, and a number of hypotheses led to the measurement of a form of beryllium, and also iridium, in the clay.

For reasons relating to the dynamic formation of the Earth's iron-rich core, iridium is effectively absent from the Earth's crust. But it is deposited by extraterrestrial dust at a relatively constant rate, so that concentrations in the K/T clay might be an indicator of the rate at which the clay accumulated. Rather than showing a constant low concentration suggesting slow accumulation, however, the measured concentrations of iridium showed a spike in that clay that defied prediction. There was no known process that could account for so much iridium in that layer.

Meteors and asteroids in our solar system were formed at the same time as the Earth and of the same materials but are too small to support the gravity- and heat-driven processes that removed iridium from the surface of our planet down into the core. As a result, meteors and asteroids retain an elevated concentration of iridium.

This suggested to Walter and Lewis Alvarez, and a few colleagues, that perhaps a meteor strike had caused both the clay formation at the K/T boundary and the mass extinction.

Uniformitarianism was challenged, and catastrophism was back in play.

That idea represented a radical challenge not only to existing paradigms on the extinction of the dinosaurs but also to the most fundamental paradigm in geology: uniformitarianism. As expected, this challenge generated significant resistance and debate. Following the first paper presenting the iridium data, additional research substantiated that there was indeed a global occurrence of the K/T clay layer, all deposited at the same time, and all enriched in iridium. A meteor strike seemed the most logical cause, but Alvarez emphasizes that "scientific hypotheses are tested in the crucible of intensely skeptical criticism." And this was no exception.

Throughout his book, Alvarez emphasizes that this is as it should be—that this is how major advances in science happen—not by a sudden discovery and immediate acceptance, but by the long and arduous process of conflicting hypotheses, new measurements, and scientific debate. This is the chaotic phase in a scientific revolution. One measure of the impact of this new paradigm was that, while the idea of the K/T extinction was nearly absent in the scientific literature before 1980, the following decade saw more than two thousand papers published on this idea.

The full story involves a systematic and sequential exploration of geological evidence, including the discovery and distribution of unusual minerals (shocked quartz and tektites), of massive marine sediments now explained by an unimaginable tsunami generated by the meteor strike, and finally the identification and location of the crater resulting from the impact. This final step involved geologists from the Mexican oil company Pemex, who had drilled in the area of the strike and had data unavailable to academic scientists.

And evidence continues to accumulate. An article in 2019 announced the discovery of a site in North Dakota with an incredible concentration of marine fossils and associated clues that could only be explained as the site of the farthest reach of the meteor-initiated tsunami, where massive amounts of water, marine life, and other debris were deposited before the wave retreated.

Even space physics became involved once again, beyond the iridium story. There is nothing like visible evidence for an event that, up to this point, could only be imagined. In 1994, fragments of the Shoemaker-Levy comet were predicted to impact the surface of the giant gaseous planet Jupiter, and impact they did. The degree of disturbance caused by this impact supported the notion that such extraplanetary objects were still capable of massive disruption.

And even climate change, of a sort, was involved. The concept of nuclear winter had been developed as a means to show that no one could win a nuclear war, because of the amount of dust and debris that would shock the climate system for many years. The ideas and calculations associated with this concept were also applied to predict the climate impacts of the K/T meteor impact.

The final conclusion? A meteor up to 15 km in diameter struck the coastal area north of the Yucatan Peninsula, resulting in massive physical changes to the Earth and to its climate that resulted in the extinction of the dinosaurs and, incidentally, the rise of the age of mammals. Any number of dramatic visualizations of this massive collision can be found on the Web.

One event highlights how shifting paradigms can be addressed most effectively, and also captures the kind of intellectual intensity and scientific fun that can be part of the process. In a section of his book entitled "The Detectives Gather," Alvarez describes a meeting convened in 1981 to hash out the new evidence and its possible implications. Gathering at a remote ski resort in the summer, scientists from the broadest range of disciplines came together for the kind of free-wheeling discussion that scientists, often restricted by the Iron Rule of Observation and the formality of professional publication, crave. Something like a compressed version of the Stockholm Physics Society.

By the time Alvarez's book was published in 1998, the meteor strike theory was well established and generally accepted. But paradigms do not go unchallenged. Vulcanologists were aware that perhaps the largest observable series of lava flows (the Deccan Traps in India) overlapped with the K/T extinction event. After 2000, hypotheses surfaced that it was this elevated volcanic activity over millions of years, with the resulting acidification of the atmosphere and changes in climate, that were actually responsible for the extinction event. Uniformitarianism of a sort was back in the picture.

The conversation continues, and I have to tell this story. In presenting the K/T extinction and the relative merits of the meteor hypothesis versus the Deccan Traps hypothesis to my class in the early 2010s, I did venture to say that it seemed to be just too much of a co-incidence that the largest meteor strike in the past several hundred million years and the largest series of volcanic eruptions still visible on the Earth happened to occur at the same time, and at exactly opposite sides of the globe. Could the energy of the meteor impact have caused this massive eruption? One of the fun things about teaching a class like mine is that you can just throw ideas out there and get student reaction.

Like the idea of moving continents derived from looking at a globe, no doubt many others had the same idea. Still, it was encouraging to see an article, published in 2015, that suggested just this linkage. The energy transferred to the Earth by the meteor strike could have driven an increase in the volcanic activity.

So while I was wrong about El Niño ending the California drought, maybe this little bit of speculation that links the meteor and the Deccan Traps into a single unified view might prove to be correct. Stay tuned—the time line for this topic might not be quite complete!

*

These three examples of fundamental paradigm shifts in Earth system science demonstrate the wide range of traditional research areas in-volved in the solution, and capture the long process of scientific give-and-take required to reach what might be a final answer. Do the same concepts and extended time lines apply when science is used to ad-dress pressing environmental concerns?

9

The Stages of Environmental Grief

The last meeting of my class for first-year environmental science students is entitled "Problems Solved or Not?" Having mastered all of the course content to that point, the students are asked to research several different environmental issues that have galvanized public opinion at one time or another. Their task is to answer these five questions:

1. What is the issue? What pollutant or process was seen as a threat?
2. How does that pollutant or process impact quality of life (health, cost, aesthetics)?
3. What research has been done on that issue?
4. What steps, if any, were taken to resolve the issue as a result of that research?
5. Were those steps effective? Has the issue been solved?

One of my primary reasons for doing this is to reinforce the idea that many of those issues have indeed been resolved through a process that involved good science used by policy makers to deliver political and economic solutions. This is the way it is supposed to work. Students need to know this as a counterweight to some of the end-of-the-world hopelessness that young people feel about their world (that sense of doom is captured well in the opening chapter of Robin Wall Kimmerer's *Braiding Sweetgrass*). There are even Wikipedia pages on ecological grief and on climate grief. What's the message? Major problems can be, and have been, solved.

What are some examples? Well, there is no longer lead in gasoline or paint (if you wanted to spread a deadly toxin generally over a twentieth-century population, adding it to gasoline and paint would be excellent choices). The production of DDT has been greatly reduced, thanks to Rachel Carson. The killer fogs of 1950s London are (nearly) gone. The concentration of smog in the Los Angeles basin has been reduced by about 80 percent since the 1950s (growing up there I never saw the mountains off to the east of the city, and school children were often subjected to smog alerts and asked not to run or even play outside). The Cuyahoga River in Cleveland hasn't actually caught fire for several decades now. Eutrophication of rivers and lakes has been reduced in part by the simple elimination of phosphorous from laundry products.

The list of issues that have been addressed and solved or at least largely reduced or mitigated is a long one. But none of these successes was easily or quickly won. Each required time and several steps, both scientific and political.

And the steps involved also follow a pattern, which I am going to call the stages of environmental grief. This builds on the psychological pattern of human responses to major losses, especially the death of a loved one, but I'll rename those stages and provide a different final one.

In psychology, one rendering of the five stages of grief is denial, anger, bargaining, depression, and acceptance. For our discussion of the process of addressing and possibly solving environmental problems, let's call them discovery/recognition, denial, anger, understanding, and action/resolution.

That final step is the biggest refinement of the concept. The stages of psychological grief cannot return what has been lost. In environmental grief, if we get to the final stage, there is resolution of the issue, and a solution.

There are other uses of the term "environmental grief," but it is not a settled concept, and I hope this redefinition helps in understanding the role of science in addressing the three big environmental issues presented here: acid rain, the ozone hole, and the Gulf of Mexico Dead Zone.

*

If you were following environmental crisis stories in the 1970s and 1980s, acid rain was at the top of your list. Caused by emissions of sulfur and nitrogen, especially from power plants, acid rain raised concerns ranging from the death of trees, to the acidification of streams and lakes, to the premature weathering of built surfaces, including gravestones!

But acid rain, like so many other issues we have covered, has a longer history.

The industrial revolution in England was powered by the steam engine, and the steam engine was powered primarily by coal. We have seen that Svante Arrhenius was aware of the potential for industrial combustion of coal to alter climate, and there were other effects as well.

Coal is a product of the partial decomposition of organic matter produced primarily by plants. In wetlands and bogs, dead plant material is preserved by an environment that is anoxic (without oxygen). Over a period of millions of years, this organic matter can be buried and subjected to increased heat and pressure, converting it into coal (and natural gas and oil as well). So yes, coal is being produced now, but at a rate that is essentially zero in human terms, making coal (and natural gas and oil) a nonrenewable resource.

Being derived from plant material, coal contains not only the organic carbon compounds that can be burned for energy but also major plant nutrients like sulfur and nitrogen. How much of these depends on how completely the material has been transformed. Soft coal, or lignite, is less modified and contains higher concentrations of sulfur (3–7 percent). Hard coal, or anthracite, is more completely modified, has lost much of its sulfur content, and contains a much lower concentration of this element (0.2–1.2 percent).

Much of the European coal used to drive the industrial revolution was lignite, and some of the environmental impacts were recognized early on. The chemistry behind the formation of acid rain was first discovered by Robert Angus Smith in 1852 and was included in a groundbreaking book, *Air and Rain: The Beginnings of Chemical Climatology*, published in 1872.

The chemistry is fairly simple. Combustion requires oxygen, and burning of coal creates oxidized forms (or oxides) of the compounds present. The variety of carbon compounds in coal are converted to carbon dioxide (mostly). Nitrogen and sulfur are also converted to oxides of these elements. Ejected into the atmosphere by smokestacks, these oxides combine with water vapor and liquid water in the atmosphere to form sulfuric and nitric acid, two strong acids that freely donate their hydrogen ions (the determinate of acidity) to whatever form of water they encounter.

So sulfur, and to some extent nitrogen, are the conveyors of acid rain.

Having covered the basic chemistry of acid rain in my class, it is a good time to provide a historical distraction. Coal was also the primary fuel used for heating of homes in London in the late 1800s, and combustion in open fireplaces was a terribly inefficient process, generating smoke and dust as well as sulfur dioxide. In addition to driving the formation of sulfuric acid, sulfur dioxide can stimulate the formation of clouds (a process called "nucleation"). Put another way, the foggy days in old London town, captured in all those old Sherlock Holmes films, were partly the result of sulfur dioxide pollution. In class, I use a clip from an old black-and-white Holmes film to make the point. On a less humorous note, it took the London "killer fog" of

1952 to lead to the Clean Air Act of 1956 in England, which brought a reduction in this deadly form of pollution.

In the United States, acid rain has a shorter history. The Hubbard Brook Experimental Forest was established in the White Mountains of New Hampshire in the 1950s, initially as a hydrology research station of the U.S. Forest Service. In the 1960s it became the site for the Hubbard Brook Ecosystem Study, one of the first (and now one of the longest running) forest ecosystem study sites in the world. A first goal for the project was to measure the chemical balances in several watershed ecosystems located in the research forest.

Measuring those input-output balances required good numbers on the inputs to the system by rain and snow. To the surprise of the investigators, the acidity of rainfall (the concentration of hydrogen ions) at Hubbard Brook was about twenty-five times higher than would be expected in rainfall in an area like northern New Hampshire, considered to be far beyond the reach of major sources of air pollution. The discovery/recognition stage was complete, and the acid rain controversy in the United States was launched.

Let's walk through those five questions asked at the beginning of this chapter, keeping in mind the stages of environmental grief.

1. What is the issue? What pollutant or process was seen as a threat?

The pollutant was the acidity generated by reactions between oxides of sulfur and nitrogen emitted from fossil fuel combustion (especially coal) and atmospheric moisture, and the deposition of the resulting acids to forests, fields, and waterways hundreds of kilometers from the source.

Ironically, one solution to the killer fog in London was to build taller smokestacks, to inject emissions higher into the atmosphere so they would be carried away from urban centers. The "away" turned out to be remote areas up to hundreds of kilometers downwind. As in Europe, taller stacks in the United States injected emissions into the upper atmosphere, resulting in long-distance transport.

2. How does that pollutant or process impact quality of life (health, cost, aesthetics)?

At the same time that the acidity of precipitation in rural New Hampshire was being documented, it was discovered that spruce trees at high elevations in mountains across northern New York and New England were declining and dying at an alarming rate. Was it acid rain? In addition, concerns were raised about the effects of acidified waters on the food chain and fish health in streams and lakes. Potential damage to buildings and other built surfaces was also considered.

It should be mentioned here that acid rain was even more intense in western Europe, with a higher concentration of industrial activity and an early dependence on lignite as an energy source. I was involved in nitrogen deposition research at the Harvard Forest in central Massachusetts and on one occasion hosted a delegation of scientists from Germany. When I mentioned how much nitrogen came down in the rain and snow at this site, one of our German colleagues exclaimed, "But that is pristine!" In central Europe deposition rates could be ten times what they were in New England.

3. What research has been done on that issue?

As the 1970s drew to a close in the United States, there was intense political pressure to begin to regulate power plant emissions to reduce acid rain. A decision point arose—regulate or study? As the debate grew heated, it was clear we were in the denial and anger stages.

Arguments back and forth were the same we repeatedly see at this stage. Those most likely to have to bend to any new regulations say we do not know enough, and more research is needed. Those who want to take action now would argue we do know enough. The decision in 1980 was to study, and an intensive program of measurement of both deposition and its effects was initiated. The understanding stage was extended.

The National Acid Precipitation Assessment Program (NAPAP) was the umbrella for one of the largest and most intensive studies of the biogeochemistry (yes, that new discipline included the three fields present in the name, and physics as well) of natural terrestrial and aquatic ecosystems. Out of this study came greatly increased understanding of the cycling of nitrogen and sulfur, the chemistry of acidic waters, and interactions with heavy metals, especially aluminum.

This increased understanding allowed a focus on key indicators that could be addressed by policy. In Europe, the concept of critical load was defined in terms of the deposition of acidifying compounds relative to the ability of soils to counteract acidification. Maps of exceedances, those locations where deposition exceeded soil capacity, helped guide policy decisions to reduce deposition. In the United States, the role of deposition in acidifying surface waters, and especially the dissolving of aluminum into those waters and its toxicity to fish, became central.

An aside here on the role of academic freedom and the importance of independent universities and research groups in bringing an issue to the solution stage. I was at the University of Virginia when a major grant was awarded by an industry group for research on acid rain. A condition of the award was that the funder would have the right to review all findings before publication and would essentially hold veto power on the results. The lead investigator on this grant, Jim Galloway, and the university administration recognized infringement of academic freedom when they saw it, and they had that clause removed from the contract. The work was funded and proceeded.

4. What steps, if any, were taken to resolve the issue as a result of that research?

Here we encounter the interface between science and policy. Even as the evidence for the impacts of acid rain was accumulating, resistance to regulation remained strong. Pioneering researchers such as Gene Likens were to appear frequently at congressional hearings to present the science. The reception was not always cordial. The tale of the political interactions behind the decisions made in this period is beyond the scope of this book but will contain few surprises.

NAPAP had a ten-year time frame, and as 1990 approached, the political climate was different from what it was in 1980, and, it would be nice to think, the scientific evidence had become convincing. In that year, Congress passed the Clean Air Act Amendments that, among other criteria, set standards for the emission of sulfur (and to some extent nitrogen) from point sources like power plants. A unique provision of this bill was that reductions were to occur not through direct regulation of individual sources by government agencies but by setting regional goals and allowing the polluters to trade allowances to emit sulfur. This cap-and-trade approach proved to be highly effective in bringing the active participation of the emitters into the process and allowing the marketplace to determine where reductions would be realized most efficiently.

5. Were those steps effective? Has the issue been solved?

The policy was effective. Reductions in emissions resulted partly by switching from soft to hard coal, partly by upgrades to existing plants, and also by converting many power plants from coal to natural gas. Details of technology and policy are beyond the scope of our story here, but the overall impact has been a reduction in sulfur depo-

Three-Year Wet Sulfate Deposition

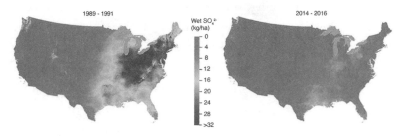

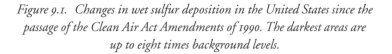

Figure 9.1. Changes in wet sulfur deposition in the United States since the passage of the Clean Air Act Amendments of 1990. The darkest areas are up to eight times background levels.

sition in the eastern United States by about 75 percent (figure 9.1). Similar actions in Europe have resulted in similar reductions.

For my students the lesson was this: good science, even if it took ten years to complete, played an important role in the eventual solution of the acid rain problem.

In terms of the stages of environmental grief, this process had moved through to the resolution stage.

A final story here. I was peripherally engaged in acid rain research, with a focus on nitrogen, and had received some funding from EPA (Environmental Protection Agency) for my work. After 1990, a mission-directed agency like EPA shifted focus away from acid rain, as that had been addressed by the Clean Air Act Amendments. I remember the day that I called the office at EPA that had dealt with acid rain to ask a question about continuing work. The person on the other end answered the phone with the greeting, "Climate change!" At least at that point, EPA had moved on to the next big issue.

*

If acid rain was the premier environmental issue of the 1970s and 1980s, the ozone hole was a solid second.

The chemistry of rare gases in the atmosphere is a unique, and uniquely challenging, area of environmental research. The complex reactions among the compounds involved, the concentrations at which they occur, and the huge variation in temperature, pressure, and radiation intensity from the Earth's surface to the top of the stratosphere (about 50 km up) all confer some of that unique status.

Centered about 25 km up in the atmosphere is a region (known now as the ozone layer) enriched in ozone, a molecule consisting of three atoms of oxygen. This concentration results from intense solar radiation at this height that splits normal molecules of oxygen (with two atoms, about 21 percent of our atmosphere) into single atoms, some of which combine with other two-atom molecules to form ozone (O_3). Yes, this is the same molecule that is a major component of smog when it occurs at the surface. It is an important pollutant at ground level because it is highly reactive, especially with soft tissues like lungs and the internal cell surfaces of leaves.

As a reactive compound, ozone is short lived in the atmosphere and must be recreated continuously. This is a good feature for controlling ground-level ozone but a problem for the ozone layer in the upper atmosphere. Ozone is extremely effective at absorbing the highest energy, shortest wavelength forms of radiation emitted by our sun. Without the ozone layer, additional ultraviolet radiation would reach the surface, damaging the skin, eyes, and other tissues of humans and other animals, and also generating genetic abnormalities.

Two ironies might help highlight the complex story of ozone. First, it is the high-energy, ultraviolet radiation that splits the oxygen atoms in the upper atmosphere and leads to the creation of ozone, which then absorbs that same high-energy radiation, protecting those of us at the surface. Second, growing up in Los Angeles with high levels of ozone at ground level, were we protected from ultraviolet radiation in that sunny locale but maybe more likely to develop damaged lungs?

Frank Rowland and his Ph.D. student Mario Molina were early leaders in studying the chemistry of the upper atmosphere. An early hypothesis that the emissions of oxides of nitrogen by supersonic transport (SST) jets could deplete the ozone layer had been dismissed, but Rowland and Molina hypothesized that recently invented chlorofluorocarbons (CFCs) could be much more effective in this regard. In essence, the same high-intensity radiation that creates the ozone layer could also split the CFCs, releasing individual atoms of chlorine. The chemistry is complicated, but ultimately chlorine disrupts the ozone creation/reaction system and can result in a reduced concentration of this form of oxygen.

The discovery/recognition stage was complete.

So let's consider the five questions:

1. What is the issue? What pollutant or process was seen as a threat?

CFCs produced and released at ground level were inert in the lower atmosphere and would eventually migrate to the upper atmosphere. There, intense radiation would separate a chlorine atom from the CFCs, and that would disrupt the ozone cycle, leading to the depletion of this gas in the ozone layer.

2. How does that pollutant or process impact quality of life (health, cost, aesthetics)?

The ability of ozone to absorb high-energy radiation protects living organisms on the surface from harmful ultraviolet radiation. Depleting this layer could result in important health issues for humans and other animals.

3. What research has been done on that issue?

Research proceeded rapidly. As sunlight is required for the formation of ozone, the depletion of ozone over the south pole in winter was a natural occurrence. Adding chlorine accentuated the rate of disappearance of ozone in the Antarctic winter, creating what became known as the ozone hole.

How did we measure the size of the hole? The Nimbus 7 satellite, launched on October 24, 1978, had an instrument named the Total Ozone Mapping Spectrometer (TOMS), which could return information on the total amount of ozone in the air column over the locations seen by the satellite. In this case, the understanding of the chemistry and the generation of the depletion hypothesis co-occurred with the primary method for gathering data. TOMS produced dramatic images of the ozone hole and allowed continuous monitoring of the size of the hole and total ozone concentration over Antarctica (figure 9.2).

4. What steps, if any, were taken to resolve the issue as a result of that research?

The time line is very short. Research on the impact of CFCs and the measurement of the size of the ozone hole led almost immediately to calls for regulation. The commercial interests allied with the production and use of CFCs generated predictable resistance to regulation. In this case, the denial, anger, and understanding stages proceeded in tandem.

Chlorofluorocarbons were used primarily as the refrigerant in air conditioners and refrigerators but were also used as propellants in spray cans and for other industrial purposes. One of the first public reactions to the ozone hole story was a negative view of spray cans, and efforts were made to replace them. Not everyone agreed. One personal encounter with a clerk using a spray can in a store near our

1979   1998

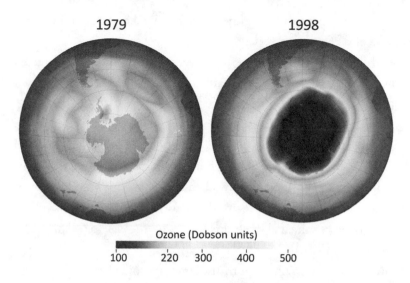

Ozone (Dobson units)

100   220   300   400   500

Figure 9.2. Changes in total ozone content of the atmosphere over Antarctica from 1979 to 1998, showing the extent of the midwinter ozone hole. The darker area in the center of the right-hand image represents lower ozone concentrations.

home included a suggestion that maybe such things shouldn't be used anymore. The clerk cheerily responded, "But I only use it indoors!"

The global community, however, recognizing the potential danger posed by CFCs, gathered in Montreal in 1987 and produced what is known as the Montreal Protocol, effectively banning the production of CFCs and related substances worldwide.

5. Were those steps effective? Has the issue been solved?

Chlorofluorocarbons are inert in the lower atmosphere and take a long time to migrate up to the upper atmosphere. Because of this characteristic, the concentration within the ozone layer changes slowly. Concentrations of certain CFCs in the ozone layer have declined, and the size of the ozone hole has been stabilized and may be decreasing. Because of this time delay, it will take decades for the impacts of CFCs to be reversed.

In addition, monitoring efforts over the past decade have detected increased CFC emissions, probably originating in China, which will forestall any gains.

Still, in terms of our stages framework, the time from recognition to understanding and resolution is very short. Eventually, this should rank as a solved problem, although it is unclear what the environmental impacts might be of the new chemicals that have been developed to take on the functions formerly filled by CFCs. In terms of our stages of environmental grief, this issue is on the way to resolution.

*

This has all been a pretty optimistic picture so far. In chapter 8 and this chapter, science is ascendant. Major questions in Earth system science have been settled through the active give-and-take of ideas and measurements driven by Strevens's Iron Rule—the polyps have been adding to the superstructure of the scientific reef. Several major environmental issues have been resolved, eventually, by the interaction between good science and effective policy.

This is not to say that all the major issues have been resolved. Climate change is the topic for chapters 10–12, but there are others as well.

Our third environmental issue deals with providing food for the growing human population. I will start that discussion by presenting the dichotomy between Malthusians and cornucopians. You will see why in a minute.

Malthusians build on the ponderings of the late eighteenth- and early nineteenth-century cleric Thomas Malthus, who stated that human population growth will necessarily outstrip the ability of the Earth to feed that growth, and that the ultimate limitation on the human population is misery—death by starvation, disease, and warfare. Malthus is sometimes called the founder of modern economic theory, which may be one reason economics has sometimes been called the dismal science.

In contrast, cornucopians posit that whatever challenges confront human society will be solved through technological and social innovation. It is unknown in advance what that innovation will be, but the concept is that there will always be one—that human ingenuity knows no bounds.

You can trace the evolution and examples of both of these lines of thought in many books and essays, but here is one example of a cornucopian solution of major proportions, and its resulting environmental impact.

By the end of the nineteenth century, Malthus seemed to have the upper hand. A pronouncement from the British Academy of Sciences proclaimed, "England and all civilized nations stand in deadly peril. As mouths multiply, food sources dwindle."

The problem was a lack of fertilizers to support the growth of food crops beyond what could be sustained by recycling manures. As this crisis developed in the second half of the nineteenth century, unique and nonrenewable sources were being ruthlessly exploited. The story of the short-lived commodity market for seabird guano mined from dry tropical islands is a sorry tale of short-term profits and much human suffering. And the resource, having accumulated on these remote islands for centuries, was essentially a nonrenewable one. From about 1840 to about 1880, guano was a hot commodity, and then it was essentially gone.

The follow-up to the dire warning above was "It is the chemist who must come to the rescue." A call for a cornucopian intervention. That intervention came to pass, and chemists were indeed the catalysts.

But let's set the context for this cornucopian story with two major ironies around the role of nitrogen in life and death.

The first is that while we are all awash in an atmosphere that is 78 percent nitrogen gas, most plants, and entire ecosystems, tend to be limited in growth by shortages of that element.

Nitrogen gas is composed of two atoms of the element linked by a triple bond. That type of bond is very difficult to break, so nitrogen gas is very unreactive—a major reason it has come to dominate our atmosphere. This tranquil gas, so named by Sam Kean in *Caesar's Last Breath*, has brought stability to the Earth's atmosphere over geologic time, much to the benefit of evolving life, but presents a form of nitrogen largely inaccessible to living things.

Because of its prevalence in proteins and genetic materials (such as DNA), nitrogen constitutes about 3 percent of living biomass. It is essential in the right ratio to carbon and other elements for balanced growth.

Before 1900, the only processes that could fix nitrogen from the atmosphere into forms available to biology were lightning (a minor source) and certain microbes with the ability to carry out this transformation using the energy from photosynthesis or derived from existing chemicals and organic compounds.

Some higher plants, especially legumes, such as table beans, soybeans, and some others, had evolved a symbiotic relationship with microbes to allow nitrogen fixation, as it is called, in their roots, trading energy from photosynthesis for the nitrogen made available by the symbionts. Rotating nitrogen-hungry crops like corn with a nitrogen-fixing species like soybeans has long been a way to attempt to increase the availability of nitrogen, but by 1900 it was not sufficient to solve the Malthusian dilemma.

The second irony is in the human uses of nitrogen, for this element is crucial both for crop production and for munitions, at least as the technology of warfare stood in 1900 (the N in TNT stands for a nitrogen component in that explosive). Life and death from the same element.

The full story of this cornucopian intervention involves the inventive genius of Fritz Haber and the commercial genius of Carl Bosch, two German scientists, and is well told by Kean in *Caesar's Last Breath*.

In 1905, Fritz Haber, a German scientist/engineer who had drifted among several different projects, was asked by an Austrian company to look into ways to create ammonia gas from the abundant supply of unreactive nitrogen gas in the atmosphere. He eventually succeeded in developing a high-temperature, high-pressure process for cracking the nitrogen triple bond and making ammonia as a first step in the production of both fertilizers and munitions. Working then with Carl Bosch, these two engineered and commercialized a method for large-scale fixation of nitrogen by this technique, which is still known as the Haber-Bosch process.

Both Haber and Bosch earned Nobel Prizes in Chemistry, Haber in 1918 for his work on producing ammonia and Bosch in 1931 for general inventions of chemical methods employing high pressure. As Kean says, it would be nice if the story ended there, but it does not. The politics and technology of World War I led Haber to be a prime mover in the development and use of chlorine and other gases for chemical warfare, which earned him the enmity of the world. He may be the only Nobel laureate who might have been, but in the end was not, tried as a war criminal. Both Haber and Bosch had ambivalent dealings with the Nazis in the 1930s, further clouding their legacies.

But politics and warfare are not our main concern here.

Invention of the Haber-Bosch process revolutionized agriculture. Suddenly, it was possible to produce nitrogen fertilizers in essentially unlimited quantities. It could be argued that this invention also made possible a second cornucopian innovation in agriculture, the green revolution, or the intensification of agriculture through the combination of genetically enhanced growth potential, especially of rice and wheat, matched with the provision of fertilizers and other chemicals. Intensification and industrialization of agriculture has as many detractors as proponents, but the green revolution did forestall famine in countries from India to Indonesia, and it kept Malthus at bay for at least a few more decades. The green revolution would not have happened without nitrogen fertilizer from the Haber-Bosch process.

As a high-energy process, the production of 175 million tons of nitrogen fertilizers in the current world each year consumes about 1 percent of the total energy production globally, and so generates a proportional amount of carbon dioxide. But for that energy investment, the nitrogen from the Haber-Bosch process essentially doubles global food production. Put another way, without Haber-Bosch, global agricultural production might only be able to support a human population of three to four billion. Haber-Bosch is truly a cornucopian response to a Malthusian dilemma, forestalling misery as the limiting factor for the number of people on Earth.

And now for the environmental impact of this cornucopian innovation.

Where does that 175 million tons of nitrogen fertilizer go? As one example, about seven to eight million tons is applied to farmland in the Mississippi River watershed, fostering huge increases in increased crop production, especially of corn.

But, in very oversimplified terms, fertilizer applications are subject to the law of diminishing returns. As fertilizer applications go up, crop growth goes up—but at a diminishing rate. If fertilizer is relatively cheap and the crop relatively valuable, it pays to continue to add fertilizer to increase yield, but as a smaller fraction of added fertilizer is incorporated into the crop, a larger fraction remains in the soil, at least for a while.

Excess nitrogen fertilizer can be converted back to nitrogen gas (a relatively neutral process) or to other gases containing nitrogen, some of which are also greenhouse gases. Much of it will be carried away by rainfall into streams and rivers, including the Mississippi. An EPA study group suggests that 60 percent of the nitrogen reaching the Mississippi delta is derived from agricultural sources.

So excess nitrogen in the Mississippi River watershed washes downstream and into the Gulf of Mexico—and here is where we encounter the Dead Zone.

Eutrophication is the term applied to the response of water bodies to excess nutrient loading. Nutrient-rich waters support algal blooms—rapid growth of algae, leading to dense concentrations of their cells easily visible to the eye, and occasionally toxic in themselves. As the algae turn over and die, the decomposition of their biomass takes up all the oxygen dissolved in the water column, and the water becomes anoxic—concentrations of oxygen too low to support fish life, for example.

And this is what has happened at the mouth of the Mississippi River.

So let's answer our five questions:

1. What is the issue? What pollutant or process was seen as a threat?

Excess fertilizer applications and other forms of nitrogen in the Mississippi watershed lead to high nutrient concentrations beyond that river's estuary.

2. How does that pollutant or process impact quality of life (health, cost, aesthetics)?

Excess nutrient loading leads to eutrophication of the waters in the adjacent Gulf of Mexico, leading in turn to the depletion

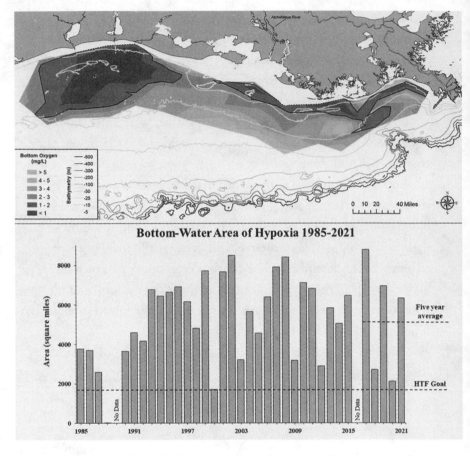

Figure 9.3. The Gulf of Mexico Dead Zone in space and time. Top: oxygen concentrations range from more than five parts per million (white, near normal) to fewer than one (the darkest shade). Bottom: the change in the maximum size of the Dead Zone each year since 1985.

of oxygen in these waters and the creation of the Dead Zone. The most recent five-year average is more than 5,000 square miles (about 13,000 km^2) of anoxic waters extending in the direction that currents flow in the Gulf (figure 9.3).

The size and extent of the Dead Zone change year to year, depending on weather conditions. Hurricanes can be especially effective at distributing the nutrient load over a wider area and reducing the size of the Zone. The discovery/recognition stage is complete.

3. What research has been done on that issue?

The process of eutrophication is well known and has been for many years. In terms of the Dead Zone itself, monitoring of the size of the Zone appears to have begun around 1985 (figure 9.3). Data accumulated related to this process include application rates of fertilizer across the region. The understanding stage is well advanced.

4. What steps, if any, were taken to resolve the issue as a result of that research?

According to the EPA website (Hypoxia Task Force) that hosts information on the program, a task force was formed in 1997 to study the problem. In 2001, the task force released a national strategy to reduce Gulf hypoxia based on an integrated assessment of the available science. A revised plan was released in 2008. At least three federal agencies—EPA, NOAA, and the U.S. Geological Survey—are involved in the process, but, under the current plan, nutrient management actions are essentially the responsibility of individual states.

5. Were those steps effective? Has the issue been solved?

No. Figure 9.3 illustrates that the extent of the Dead Zone varies widely year to year, but there is no discernible downward trend. The average size of the Zone over the past five years is about 2.5 times the stated goal of the task force.

In terms of our stages of environmental grief, discovery and understanding are well advanced, but there seems to be little activity in the denial and anger stages, because this topic does not often rise to the level of other environmental issues. Steps toward the resolution stage have not been effective.

*

At the end of this "Solved or Not?" class exercise I ask the students to say which issues have been solved and which have not. They have no trouble dealing with the statistics related to each issue, agreeing, for example, that smog in the L.A. basin has declined by 80 percent. On whether the issue is solved, however, there is often a divergence of opinion—and that offers a teachable moment. Whether the ozone concentrations in the L.A. basin are low enough is a policy decision, and beyond the scope of that class. Science presents the facts. The policy/political process needs to weigh those facts in the light of conflicting values.

So room 3 in our tour has shown that scientific progress on both the big questions in Earth system science and major environmental issues tend to follow similar patterns, captured in this chapter as the stages of environmental grief. Time lines from discovery of an issue to resolution can vary widely. For some issues, such as the Gulf of Mexico Dead Zone, there has been no resolution. Where do we stand with climate change in terms of these stages?

ROOM FOUR
Climate Change
THE DATA, THE MODELS, AND OUR FUTURE

10

The Discovery (and Rediscovery)
of Climate Change

Does climate change fit the same best-case scenario I have presented so far, where good science feeds good policy decisions and the
problem is solved? Does the same pattern from initial discovery to
ultimate solution pertain? Just where are we in the sequence presented
in chapters 8 and 9?

There are two parts to this story and this chapter. The first part
briefly retells the history of the concept and basic understanding of
climate change. The second presents the measurements that let us assess the question of state and change in the climate system up to the
current time.

As we have seen with most other issues, the basics of climate
change and the role of greenhouse gases have been known for a long
time, and there are a number of excellent, detailed presentations of
this history. Spencer Weart's *Discovery of Global Warming* presents the
story from the earliest recognition of the ability of certain gases to
absorb longwave or heat radiation through to the ongoing process of
the Intergovernmental Panel on Climate Change (IPCC). In *Thin Ice*,
Mark Bowen wraps adventure stories of collecting ice on tropical
mountaintops around a clear and concise history of the development
of the science of climate change. James Fleming's *Historical Perspectives on Climate Change* is an older volume that leans more heavily on
European sources and discoveries.

The standard story begins with Joseph Fourier, a French mathematician first and physicist second who developed many early concepts in the analysis of numerical series and in differential equations. He is the first in a series of brilliant scientists for whom thoughts about the temperature of the Earth formed an interesting intellectual sideline. Fourier calculated that, given the amount of radiant energy received from the sun, the Earth should be much cooler than it is. He is credited with being among the first to suggest, in the 1820s, that the atmosphere might play a role in determining the heat balance of the planet. His work is sometimes paired with that of Claude Pouillet, who also speculated on the role of the atmosphere in modifying global temperatures.

In the late 1850s and 1860s, John Tyndall, who combined a passion for hiking the Alps, and a resulting interest in climate and glaciers, with curiosity about the interaction of gases with different forms of light, developed a method for measuring the ability of different gases to absorb radiation in different wavelengths. He discovered that some gases that were transparent in the wavelengths we can see (visible light) are opaque, or very good at absorbing the energy contained in longwave, infrared, or "heat" radiation wavelengths. He determined that water vapor was the strongest absorber of heat radiation in the atmosphere, and also determined that carbon dioxide and methane were active absorbers. In contrast, he discovered that nitrogen (as N_2) and oxygen (as O_2), which together account for about 99 percent of the atmosphere, were transparent to both visible and infrared wavelengths.

We revisit now, again, the Stockholm Physics Society, and Arrhenius, Högbom, Bjerknes, Pettersson, and others and their discussions of the newly discovered ice ages, Tyndall's findings of the absorptive properties of trace gases, and the increased understanding of the global carbon cycle. Chapter 1 tells the story of this group and how it led Arrhenius, in 1896, to complete that first set of year-long hand calculations on the role of carbon dioxide in altering global temperatures. He occupies the third position in our time line, and we can add him to the list of brilliant thinkers for whom climate change was an interesting sidelight to a distinguished career.

The story then goes dormant for some time. Based on what is now seen as incomplete information on the fine-scale spectra or range of wavelengths involved in the absorption of longwave radiation, and on a generally skeptical view of the ability of humans to alter something as big and robust as the global climate system, Arrhenius's results fell into disfavor. As Weart puts it, there was a continuing belief in the balance of nature—that the climate system and other natural systems were self-correcting.

Our next step in the time line belongs to Guy Callendar, an English engineer and another in the continuing line of part-time climate change researchers who have undertaken massive amounts of data collection and computation. In this case, Callendar's primary career focused on applied energy systems, from steam engines to batteries to fuel cells. As an amateur climatologist, Callendar used existing temperature data from around the world to calculate, in 1938, what he determined was a 0.5°C increase in global temperature between 1890 and 1935. Weart credits an emphasis on North American and European data sources for this relatively high number.

There are good reasons climate change was not at the top of the world's agenda in the first half of the twentieth century. Convulsed by two world wars and a major economic depression, the global community experienced huge disruptions leading to unprecedented waves of immigration, a new social order, and major technological advances, not the least of which was the dawn of the age of nuclear weapons and with it the potential for global annihilation. The first half of the twentieth century stands as the bloodiest in human history.

One of the earliest postwar voices on global change was Rachel Carson, who, before publishing *Silent Spring*, declared in *The Sea around Us* that many species of North American birds had expanded their ranges to the north, as had the common codfish. As quoted by Bowen in *Thin Ice*, Carson noted that "a definite change in arctic climate set in about 1900, [and] that it became astonishingly marked about 1930." She did not connect this change with greenhouse gases.

Gilbert Plass comes next, and the story will have some familiar components. Plass was employed as a physicist and engineer by corporations such as Ford and Lockheed and was in and out of academia several times. Bowen says that it was during a sabbatical at the University of Michigan in 1954–1955 that Plass used an advanced understanding of the physics of infrared radiation in one of the first computer models of the atmosphere to predict that a doubling of carbon dioxide would warm the planet by 3.6°C. He also predicted that carbon dioxide levels would be 30 percent higher and that the planet would be about 1°C warmer in 2000 than in 1900. Those predictions turned out to be astonishingly accurate.

Bowen concludes that Plass's work was a major milestone, in that he was successful in convincing mainstream scientists that the relationship between carbon dioxide and climate was based on solid principles. Plass also apparently initiated the movement into ever-increasingly complex models of the atmosphere for both weather and climate prediction. What Plass could not say for certain was just how much the carbon dioxide content of the atmosphere had increased since 1900, because no long-term measurements existed.

At this point the history becomes infinitely more complex, and actually passes, as Weart reports, from history to something more like journalism, which has been called "a first rough draft of history." Before considering the data, I should introduce some background information on the radiant energy received from the sun and the role of different gases in capturing that energy and retaining it in the atmosphere (the greenhouse effect).

The quality of light energy is described in terms of the amount of energy received at different wavelengths. Perhaps the simplest way to visualize wavelengths is with a prism, a triangular column of glass or plastic that refracts or divides white light (as we see it) into the constituent colors sometimes summarized as ROYGBV (red, orange, yellow, green, blue, and violet). A rainbow is also the product of refracted light, with the passage of sunlight through water molecules in a misty atmosphere acting like a refracting prism. Because of the required angle of refraction, rainbows will always appear with the sun at one's back and at an angle of 42 degrees between the eye and the water droplets in the atmosphere, generating the arc shape.

Radiant energy is carried by photons, a type of elementary physical particle that has a somewhat miraculous set of characteristics. Photons are both particle and wave, and they transmit energy while having no mass (and that is as far as we are going to go into particle physics). For wavelength, picture waves on the ocean, with different distances between peaks. Photons travel in a similar wavelike pattern, but the distances between peaks in high-energy photons are measured in nanometers. Nano means one billionth, so a nanometer is 0.000000001 meters. A human hair is about 50,000 nanometers thick, so the longest wavelength in figure 10.1, 2,500 nanometers, is about 5 percent of the width of that hair!

The Earth receives radiant energy from the sun across the full range of wavelengths (figure 10.1). Shorter wavelengths (such as ultraviolet and visible light) carry more energy per photon, but the total amount of energy at any wavelength is determined both by wavelength and by number of photons. The light gray spectrum in figure 10.1 records the total amount of energy received at the top of the atmosphere at each wavelength (the solid line around this area is a theoretical calculation of what we should receive, based on the temperature at the surface of the sun). The darker area is the amount received at the Earth's surface in the absence of clouds.

The difference between the lighter and darker areas is the energy absorbed by a wide range of compounds in the atmosphere. Some of the most important are listed in figure 10.1 near the parts of the spectrum where each is most effective at absorbing radiant energy. Those absorption spectra, as they are called, are unique to each gas. Laboratory and field instruments use the characteristic peaks in these spectra to identify and measure the concentration of gases in the atmosphere. Think of them as unique fingerprints for each gas.

The choice of gases in this figure capture two important functions: protection of living tissues from the most dangerous high-energy (ultraviolet) rays, and contribution to the greenhouse effect.

For example, ozone (O_3) is very effective at absorbing the very short-wavelength, high-energy ultraviolet radiation received from the sun at the top of the atmosphere. The potential for this high-energy radiation to damage skin, eyes, and other soft tissues is one reason the ozone layer in the upper atmosphere matters to us (see chapter 9). Sunblock lotions are designed to protect against these rays. Looking at this in another way, if we were only able to see ultraviolet radiation, the surface of the Earth would appear very dark to us, as so little of this type of radiation reaches the surface!

The greenhouse effect is all about absorption of the longer-wavelength infrared or heat radiation. I've mentioned before that water vapor (H_2O) is the strongest of these, due both to its spectrum and to its concentration in the atmosphere. Arrhenius and others since realized that a warmer atmosphere will hold more water vapor, providing a positive feedback to warming caused by the greenhouse gases we are adding.

The most important greenhouse gas altered directly by human action is carbon dioxide (CO_2), emitted primarily by the combustion

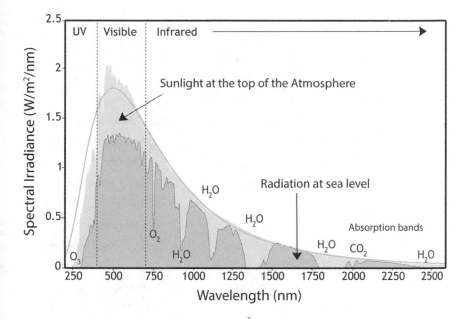

Figure 10.1. The solar spectrum, showing the amount of energy reaching the top of our atmosphere by wavelength (the light gray area) and the amount reaching the surface through a cloud-free atmosphere (the darker shaded area). The difference between these two is due to absorption by gases in the atmosphere. Regions of absorbance by ozone (O_3, at the far left), water vapor (H_2O), and carbon dioxide (CO_2) are indicated. Ozone absorbance of ultraviolet (UV) radiation protects living tissues from this harmful form of sunlight. Absorbance by water and carbon dioxide in the infrared region affects both incoming and outgoing (heat) radiation—the greenhouse gas effect. Other important greenhouse gases (nitrous oxide and methane) absorb at longer wavelengths, to the right of this graph. A nanometer (nm) is one billionth of a meter.

of fossil fuels. One absorption peak for this gas is shown in figure 10.1. Other absorption features occur at longer wavelengths, to the right of those shown in the figure. Absorption at longer wavelengths also applies to the next two major greenhouse gases: methane (CH_4) and nitrous oxide (N_2O). Nitrous oxide emissions are driven largely by agriculture, especially the use of nitrogen fertilizers. Methane emissions come both from agriculture and from the energy system. Each of these gases absorbs infrared radiation in a different part of the spectrum. Together they create an absorption spectrum for the entire atmosphere that affects the energy balance of the planet.

Let's step back for a minute and look at what I think is an amazing story told by these spectra. All of the major gaseous components of the atmosphere, including water vapor, are transparent (low absorbance), in what we call the visible wavelengths (400 to 700 nanometers). We call this light visible because it is the light our eyes can see. It is also the part of the spectrum where photosynthesis occurs. We see healthy leaves as green because they absorb mostly in the red and blue parts of the spectrum and reflect more in the green.

I find it not at all surprising that evolution has led both vision and photosynthesis to focus on the part of the spectrum that delivers the highest total energy content to the Earth's surface (figure 10.1) through this window in the absorbance spectra of major and minor atmospheric gases. Would life as we know it be possible if this window did not exist?

In contrast, the greenhouse effect is all about infrared radiation, the long-wavelength light we can't see. But there are ways for us to experience or sense infrared radiation.

Every object (you, me, your desk, the book you are reading) emits this longwave, infrared, or heat radiation. The amount emitted increases with the fourth power of temperature (very rapidly and nonlinearly) and is also affected by the type of surface.

One possibly familiar way to experience radiant heat is to sit in front of a woodstove or electric space heater with a hot surface. That surface might not change color, as we see it, but it will emit increased infrared, longwave, heat radiation as it heats up. The warmer the surface facing you, the more heat you feel, even though there is no warmed air reaching you. If you stoke the fire a little more or turn up the knob on the electric heater, and the surface of the stove or element starts to glow red, then you are receiving radiant energy in both visible and infrared wavelengths. The hotter the surface, the more radiant energy you will receive and the warmer you will feel.

There are also technologies that let you see this type of radiation, even though it is outside the visible range. Night vision binoculars sense infrared radiation and convert it into visible light. If you are a wildlife biologist who wants to track animals at night, or an adventure gamester playing a search game in the dark, any warm-blooded creature in your field of view will be warmer than the background, emit more infrared radiation, and be visible with your night vision equipment. On the other hand, this won't work if you are tracking cold-blooded reptiles; you will need a motion detector for that kind of observation!

Greenhouse gases absorb the longwave radiation emitted by the surface of the Earth. How does this affect the energy budget of the planet?

Figure 10.2 is a simplified rendering of that energy budget. The units are watts per square meter, or energy per unit area. The term "watts" here is the same as you use to select a lightbulb or read as usage on an electric bill, but let's not concentrate on the units. Rather, let's focus on the relative numbers for each arrow that track the flow of energy through the Earth system.

The lighter arrows on the left in figure 10.2 are dominated by visible light, the dark gray arrows by infrared or heat radiation. The black arrow on the far right is transfer from the surface to the atmosphere by evaporation, and the upperward movement of heated air.

Of the 340 units of energy received from the sun, one hundred are reflected directly back to space by clouds or land and water surfaces (albedo—see chapter 6). We see the Earth from space, or the moon from the Earth, because of this reflected sunlight. The remaining 240 units are either absorbed by the atmosphere (77 units) or by the

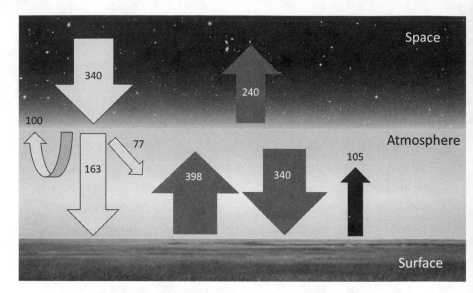

Figure 10.2. A simplified view of the energy balance of Earth's atmosphere (units are watts per square meter). Light arrows to the left represent incoming solar radiation, the darkest arrow to the right is turbulent and evaporative transfer from the surface. The arrows in the middle represent transfers of infrared or heat radiation and capture the greenhouse effect.

Earth's land and water surfaces (163 units). The most fundamental concept of any energy budget is that energy in has to equal energy out. For example, losses of energy to space (100 + 240) equals gain from the sun (340).

The heating of the land and ocean surfaces drives the internal cycling of energy through evaporation of water from oceans and lakes and through plants, as well as the convection of air from those warmed surfaces (combined in the black arrow to the right—105 units). The bigger transfer from the surface, however, in fact the biggest in the whole system, is the emission of 398 units of infrared or heat radiation by land and water. Gases and particles in the atmosphere absorb much of this energy and reradiate it back to the surface

(340 units). This recycling of energy between the surface and the atmosphere is the greenhouse effect.

To put this in perspective, the amount of energy received by the surface of the planet as infrared radiation from the atmosphere (340 units) is more than twice the amount received at the surface as visible light from the sun (163). One can see how both Fourier and Tyndall imagined a much colder Earth without the conditioning effects of the atmosphere.

As we add greenhouse gases to the atmosphere, they absorb more of the infrared radiation emitted by the surface. In effect, this increases the recycling of heat energy between the surface and the atmosphere, increasing the temperature of both.

With this background, let's return to the telling of the climate change story, which now switches from the historical sequence to the measurements that convey recent changes in the state of the climate system.

In classes or in public presentations, I like to present the Keeling Curve as the environmental icon of the climate change era. I usually show this curve (figure 10.3, top) first without any numbers or legends, just the sawtooth pattern, and ask if anyone knows what it is. If you're familiar with this image, it will be clear what it represents, even without the labels. This is an image that should be familiar to everyone, but only rarely do I get a positive response.

Beginning in the 1950s, Charles Keeling was driven to measure carbon dioxide in the atmosphere with a precision not possible up to that time. He established an observatory on top of Mauna Loa, a volcano on the big island of Hawaii, and began collecting data for this environmental icon.

Continuous and accurate measurements for more than sixty years show a steady increase in the carbon dioxide concentration (figure 10.3, top). The sawtooth pattern in the data results from the fact that the measurements are precise enough to capture the impacts of seasonality in biological activity in the northern hemisphere

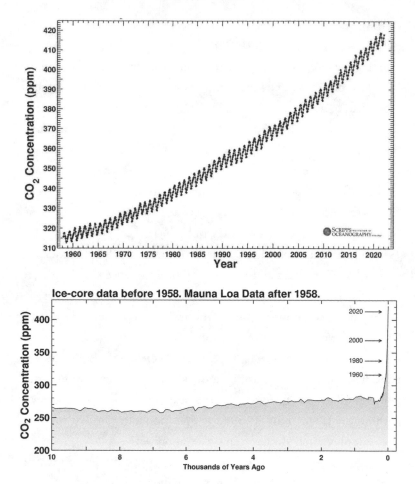

Figure 10.3. *Changes in the carbon dioxide concentration of the atmosphere over different time scales; (top) the Keeling Curve measured at Mauna Loa, Hawaii; (bottom) a ten-thousand-year time line reconstructed from ice cores.*

(dominated by photosynthesis in the summer and by decomposition in the winter). The climate change story, however, is the continuous increase in concentration over the whole period of measurement. This is the data set that Plass did not have for his detailed calculations.

Intriguing as this data set is, can we be sure it is not just capturing one part of a longer-term cyclical trend? This is always the key question in climate change research, and to answer that question we always need a longer record.

This timeline is extended back ten thousand years in figure 10.3b, using concentrations captured in ice cores (chapter 6). This version of the data set is often referred to as "the hockey stick," because it is level for most of this period (the handle) before veering dramatically upward (the blade). Before the industrial age, carbon dioxide concentrations never exceeded three hundred parts per million. We are now well over four hundred parts per million.

That time line can be extended back through eight hundred thousand years and several cycles of change in both temperature and carbon dioxide across the ice ages (see figure 6.2), and again, the concentration of this key greenhouse gas has never exceeded three hundred parts per million. Combining ice core and proxy data of the type presented in chapter 7, it seems safe to say that carbon dioxide concentrations have not been as high as they are now for at least the past twenty million years.

Relative to the brief span of human history, we are in uncharted waters, atmospherically.

And what has caused this recent increase in carbon dioxide? Primarily the combustion of fossil fuels—coal, oil, and natural gas. Depending on the time frame, about 50 percent of those total historical emissions remain in the atmosphere. Where are the rest? Primarily in the oceans, which chemically absorb a significant fraction of emitted carbon dioxide, a fact that was known to Arrhenius and Högbom in Stockholm in the late 1800s.

All of our story to this point would lead us to expect that the average temperature of the atmosphere would have increased since the 1950s. That is the case, but we are not going to discuss that yet. I've also described the atmosphere as the most dynamic, chaotic, and variable part of the climate system, subject to large and unpredictable changes. The huge data sets on temperature now available allow supportable conclusions, which I present a bit later, but there are parts of the climate system that change more slowly and continuously than the atmosphere and could provide more stable, less noisy evidence of a changing climate. You will know by now that we are talking about the oceans and ice.

Oceans are the largest reservoir of water and heat in the Earth system. There is a global ocean circulation system that connects all the major surface currents with deep ocean currents that complete the global cycle. One turn of the cycle may take more than a thousand years to complete. Heat transferred to the oceans by a warming atmosphere then may slow the increase in the temperature of the air, and the circulation system may then sink some of this heat into the lower depths, removing heat from the surface and buffering what would otherwise be a more rapid warming at the surface. One site puts the total accumulation of heat in the oceans at 90 percent of global heating since 1970, much of this at lower depths.

All fluids expand with increasing temperature, and ocean water is no exception, so while the oceans have buffered the increase in global atmospheric temperatures, the absorbed heat has led to sea level rise. Estimates of this process have been reconstructed back to 1880 and show about a 20 cm or 8 inch rise in sea level over that time period.

Even reduced rates of surface ocean warming would be expected to have an effect back on the atmosphere. Tropical cyclones, such as hurricanes in the Atlantic Ocean, draw their energy in part from warm ocean waters. The story of Hurricane Katrina in 2005 and how it strengthened as it passed over unusually warm water before striking the coast is a dramatic example, but individual storms do not tell a global climate change story. There is, however, an accumulating data set that charts the total energy embedded in all Atlantic hurricanes each year. There is huge variability year to year, but there is also a statistically significant trend showing that the average annual total hurricane energy has about doubled over the past 150 years.

The role that the oceans play in absorbing carbon dioxide also may have an important cost. Carbon dioxide dissolves in seawater by forming a weak acid—carbonic acid. The oceans are alkaline rather than acidic, so even this weak acid has increased the hydrogen ion content (or acidity) of the oceans by about 30 percent. This can have important impacts on many of the same organisms that sink carbon dioxide into the deep ocean by forming calcium carbonate (to be transformed eventually into limestone) in their shells, which then sink to the bottom. Even slight acidification of seawater can inhibit the formation of these carbonate shells, to the detriment of those species and this important carbon sink.

What does the ice tell us? NASA's Earth Observatory, at least of this writing, presents dramatic videos of the reduction in the extent of ice coverage in the Arctic Ocean over a period of forty years. Data are also available that record the change in minimum ice coverage in late summer (figure 10.4). These show a decline of almost 50 percent since 1980.

Given this trend, in what year would you predict that sea ice will disappear completely in midsummer? The simplest approach is just to put a line through all the points (statistically, of course, as in figure 10.4) and project that line to zero. The result is somewhere around 2075.

Beyond its value as an indicator of change, Arctic Sea ice has important feedback roles as well. As a first feedback, the difference in albedo between ice and water is such that this melting will increase the absorption of sunlight, accelerating warming and further melting of ice (so maybe that straight line projection in figure 10.4 isn't the best approach).

A second potential feedback involves the location of the remaining summer ice. Wind patterns in the Arctic tend to push the ice pack up against the northern coast of Greenland, providing a bit of cold

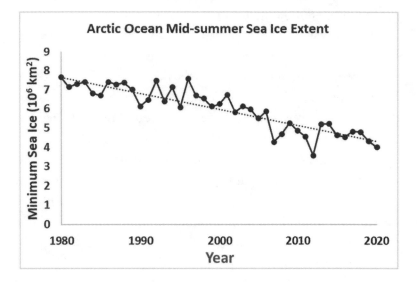

Figure 10.4. The change in the minimum (midsummer) area covered by sea ice in the Arctic Ocean over time. The dotted line is a statistical fit that can be used to predict when minimum coverage will go to zero.

insulation for that warming island. The disappearance of this ice pack may well further increase melting of Greenland ice.

Melting of sea ice, as in the Arctic Ocean, does not raise sea levels, just as melting ice cubes in your glass does not raise the level of the fluid contained. It is the melting of land-based ice that increases sea level, and indicators of the retreat of that ice abound. Measurements of glaciers worldwide capture the global reality of the kinds of trends seen in dramatic images of individual glaciers. One summary data set shows a 68 percent decline in the area of thirty-seven representative glaciers from around the world. Time-lapse images from Glacier National Park in the United States capture similar changes. Ironically, some have begun to call this national treasure the park "Formerly Known as Glacier." Similar images are available for glaciers around the world.

The big actors in terms of the loss of land-based ice are the Greenland and Antarctic ice caps. These play an outsized role in delaying the current rate of change but also convey unrelenting momentum to our climate trajectory (see chapter 12).

There is a final form of ice that is less apparent but perhaps even more important in terms of human impact and feedbacks to the climate system. Permafrost is defined as soils that are frozen year-round. About 25 percent of the land surface area of the northern hemisphere is permafrost (Russia and Canada have huge expanses of this landform). In marginal permafrost areas, the surface soil thaws in the summer, but only to a certain depth, and then refreezes in the winter. Permafrost has provided stable (if chilled) ground for human occupation and activities in the far north for millennia.

Permafrost melting has led to some dramatic images of sinkholes formed by melting permafrost devouring homes and roadways. Perhaps the most dramatic stories are recent ones out of Siberia, where the melting of permafrost apparently unlocked previously trapped reservoirs of gases that were released in explosive events, creating major craters. One of the bigger unknowns in climate change research is just how fast the global permafrost landform will melt, and also what will happen to the huge amount of carbon stored in the organic matter contained in those frozen soils. If released to the atmosphere, the extra carbon will be a strong positive feedback for global warming.

With this understanding of the role of oceans and ice in controlling the long-term changes in temperature, how have surface temperatures changed, and how has that change been distributed across the globe?

A tremendous amount of detailed work has gone into maintaining an unbiased and consistent global temperature record, and both the temporal and the spatial changes are clear (figure 10.5). In the historical record, global mean temperature has increased by more than 1°C or 1.8°F since 1980.

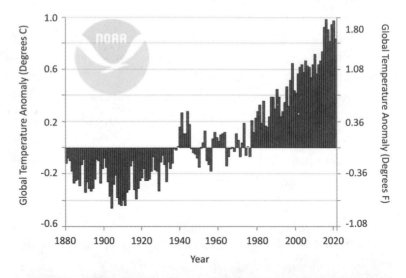

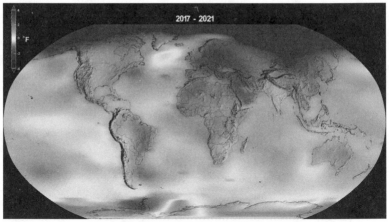

Figure 10.5. (top) The average global temperature anomaly, 1880–2020; (bottom) the distribution of change in average annual temperature, showing greater increases at higher latitudes. The darker shades denote larger increases.

In the mapped changes, just as Arrhenius and everyone since predicted, warming is more pronounced in the Arctic than at the equator. This has clear implications for the thawing of permafrost.

Precipitation trends are less clear, although for most of the United States average annual precipitation has been increasing somewhat, in keeping with what Arrhenius (and countless others since) noted: that increased temperatures would lead to increased water vapor in the atmosphere, leading to increased precipitation.

Many government agencies and private and public research organizations have contributed significant amounts of time and resources to produce the information and graphics in figure 10.5. There has also been an overarching, integrative effort to derive the scientific consensus on how the climate system works and how it might change. That effort is led by the Intergovernmental Panel on Climate Change (IPCC).

The IPCC was convened under the auspices of the United Nations in 1988. It may be ironic that this was the same year in which NASA climate scientist James Hansen made his clear and definitive statement to the U.S. Congress about the reality of climate change as a result of human activity. He described "with a high degree of confidence" that "it is happening now." Thirty-five years later, the IPCC process has come to the same conclusion.

Weart points out that the IPCC marks a departure from the process which I described for acid rain and CFCs in chapter 9, and which has been used for other pollutants as well. Instead of government panels calling on the existing expertise of scientists to explain the relevant science so that effective policies can be derived, the IPCC put scientists and policy makers together for the simultaneous process of summarizing the science and proposing policy solutions. Weart remarks that this might have been an effort to curtail some of the more radical statements that might have otherwise been generated by scientists!

The IPCC process has generated the most comprehensive summary of scientific understanding of the climate system based on a compilation of the detailed, long-term data sets used to derive that understanding. The IPCC is to climate change what NAPAP was to acid rain, but on steroids, being both international and combining science and policy. The sheer volume of the output from this process makes it difficult to summarize and present in a book like this, but it is instructive here to recap how the summary statements about climate have changed over time, before returning to the IPCC for a look at our climate future in chapter 12.

The first synthesis documents appeared in 1990 with a commitment to repeat the analysis every five years. Additional reports were released in 1995, 2001, 2007, and 2013. The science document emerging from the Sixth Assessment was released in 2021, and the full summary report early in 2022. The documents have become incredibly comprehensive, and nearly impenetrable for the general public, often running to thousands of pages. Summaries have been produced for policy makers that provide the most important conclusions.

It has been claimed that the IPCC science reports represent a consensus of scientific thought on the factors affecting climate, and I would agree that this is a valid description. Although the 1990 reports were produced by a relatively small group, the more recent cycles have included up to two thousand individuals as writers, contributors, and reviewers (full disclosure: I was a reviewer for an early IPCC cycle but have not been involved in the process for the past four). The large number of participants may help explain the exponential increase in length. The science volume alone from the Sixth Assessment runs to just under four thousand pages and includes thousands of references to technical reports. We have come a long way from the days of amateur mega-calculators such as Arrhenius and Callendar.

And what are the key findings? Although the amount of supporting data has accumulated rapidly, the fundamental conclusions have remained constant. Some of the most basic conclusions I summarize in chapter 12 have not really changed in more than a hundred years. What has changed is a continual increase in the confidence with which those conclusions can be stated.

The first report, in 1990, concluded that the absorbance of heat radiation by carbon dioxide and other gases is real, that carbon dioxide provides more than half of the greenhouse gas effect, and that global temperatures are higher than at any time in the past ten thousand years. The report said, however, that the increase of 0.3 to 0.6°C over the past one hundred years is in the range of natural climate variability and that more data would be needed to say the increase is due to greenhouse gases. Computer models of climate change played a central role in this report, and policy became blended with the science in this report, in that scenarios were generated for how emissions of carbon dioxide and other gases will change over time, depending on different policy decisions.

Those conclusions are not exactly the stuff of headlines and express the cautious nature of the IPCC process.

The 1995 assessment included a separate volume on climate science that ran to more than four hundred pages. The summary for policy makers concluded that the climate had changed over the past century and that "the balance of evidence suggests a discernible human influence on global climate." Again, hardly the stuff of blazing headlines and, again, model simulations, becoming ever more complex, also became a major vehicle for developing predictions.

By 2001, more detailed and definitive statements could be made, including that the Earth was warming as a result of human activities, especially the emission of greenhouse gases, and that changes in sea level and ice were consistent with a warming world. Even as the scientific evidence behind the conclusions became more complete and more convincing, those conclusions remained couched in numerous caveats and clear statements of uncertainty. This is classic scientific progress. Questions are raised, new data are acquired. We are deep into the understanding phase of environmental grief.

The 2007 Nobel Peace Prize was awarded to the Fourth Assessment, which was released in that year, and to one of its major proponents,

Vice President Al Gore. Although this accelerated the politicization of the process, the development and presentation of the scientific understanding continued, and statements about the changes occurring in the climate system became increasingly certain. Paraphrased, these included that warming of the climate system was unequivocal, was very likely due to the observed increase in greenhouse gas concentrations because of human activity, and was affecting many physical and biological processes globally. The report projected further increases in greenhouse gas emissions into the future, with accelerating effects on the planet.

In the Fifth Assessment, released in 2013, certainty increased even further. Statements (again paraphrased) included: Human influence on the climate system is clear. Warming of the climate system is unequivocal, and unprecedented over decades to millennia. The amounts of snow and ice have diminished, and sea level has risen. Atmospheric concentrations of carbon dioxide, methane, and nitrous oxide are higher than at any time in at least the past eight hundred thousand years.

Conclusions remain similar in the science document for the Sixth Assessment, released in 2021, and again the certainty with which they are stated has increased. This report states that it is "unequivocal" that human actions are warming all places and all processes on Earth, and that the scale and rate of recent change have been "unprecedented" over the past hundreds to thousands of years. Weather extremes are increasing. The report adds greater spatial specificity to predicted outcomes and continues the process of modeling different futures, depending on what the global community decides to do about greenhouse gases and other climate forcing factors.

The cautious, data-driven IPCC approach to climate science repeats what we have seen applied to the solution of other environmental issues. If anything, the presentation of findings has been more conservative and deliberate, perhaps due to the inclusion of policy makers in the discussion (as suggested by Weart). Although the steps in the process are the same, the scale of synthesis and presentation of the science under such an inclusive umbrella is unprecedented.

For comparison, the time from discovery of the ozone hole to the Montreal Protocol, and from the forming of NAPAP to the Clean Air Act Amendments, was around a decade. The IPCC process is now in its fourth decade.

So there is clear scientific consensus that climate change is happening and that human activities are driving that change. It has taken six rounds of synthesis and assessment to reach these definitive statements, a pace I consider to be very conservative.

NASA's James Hansen was well ahead of the process with his statement in 1988 of the certainty that the greenhouse effect had been detected and was already changing the climate. There is nothing that has been discovered since 1988 that would contradict that statement.

If there is clear scientific consensus, why has even the existence of climate change been so strongly questioned in some quarters? Although the scientific community is through the understanding stage of environmental grief, and ready to help with the resolution stage, much of the public dialogue is still in the stages marked by anger and denial. The answer to that question gets into politics rather than science, and so is beyond the scope of this book.

If we understand what is driving climate change, then where are we headed, and what is our climate future?

Predicting the future means putting some numbers into models and running them out into that future. This can be done in various ways, with widely varying levels of complexity. Before returning to our climate future in chapter 12, let's take a deeper dive into models and the modeling process, and look for consistency, but also for simplicity, such that the models are understandable and a valuable means of presenting climate science.

11

Simplifying the Explanation: How Right Was Svante?

In chapter 2, I mentioned the professional editorial I wrote in midcareer that addressed the problem of overdetermination in models. The central idea was one that has often been leveled against computer models: if you have enough variables to play with, you can make a model do whatever you want. I was lamenting a bit of loose thinking with respect to ecosystem models but also attempting to capture in the opening paragraphs the idea that trying to convince skeptics using models is an uphill battle.

I have been involved with models of ecosystems for my entire career, and what I have found is that saying "This is what the model predicts" is not very effective. That only works if the listeners are convinced that you have the science right and the conclusions follow logically from the science. That you can put code together and generate an outcome is not the most powerful part of the process.

On the other hand, in dealing with complex systems such as forests or the weather/climate energy machine, individual cause-and-effect experiments and the kinds of relationships that can be expressed in a two-dimensional graph can be misleading or even wrong if that direct effect is taken out of the context of the full system, ignoring the kinds of feedbacks and interactions presented so far in this book. Only models can include that full complexity.

There is a very large community of scientists and technicians that build, modify, and run experiments with the large, complex models described in part in chapter 2. For this group, models essentially represent an integrated working hypothesis about how the climate system works. With wide access to a limited number of community models, simulated experiments can be run that will be understandable within that community. New information can change those models, and disagreements between model predictions and measurements, or among different models, are where the science behind the models will advance.

It is almost like a computational Wikipedia—a crowdsourced, integrated understanding of how the climate system works.

So what follows is not a criticism of the modeling process or the use of models to understand weather and climate and predict our future. The models represent state-of-the-art science, but the complexity of the models and the modeling process makes the use of model results in presenting climate science to general audiences something of a "trust me" moment. I believe we need something between intuitive reasoning and models that take days to run on the biggest supercomputers available.

Are there ways to present the science using simpler, data-based explanations and relationships that capture the major drivers of climate change and build logically on the science I have presented here so far?

It was the quest for this simpler explanation, and the outcome of that quest, that got me started on writing this book. Frustrated by the complexity of climate change presentations and the willful dismissing of the issue that this made possible, I wondered if there was just a simpler way to make the point.

The classroom provided the setting for this quest. Being ready to face the bright minds in an introductory course in environmental science requires an ongoing synthesis of the overwhelming and rapidly changing amount of scientific information into a coherent story line. As one of my teaching colleagues put it, teaching a science class well and in an interesting fashion is an exercise in creative nonfiction.

In the climate part of that class we go through the steps and topics covered in chapter 10, from Tyndall and Arrhenius through to the IPCC. The data sets are compelling, but there is always a bit of magic involved in explaining the models used to make projections into the future. The models are so complex that they cannot be deciphered in a class of this kind.

So curiosity and necessity led to a search for a simpler way to present this story, and for a way that students in the class could do something interactive with climate data, and that led back to Arrhenius. I had been aware of his work in the way that most folks in the field are—an early voice that linked carbon dioxide and climate in an oversimplified way. As Mark Bowen said in *Thin Ice*, many thought Arrhenius was lucky as much as right, although Bowen opts for right, and I agree with him. That yearlong set of tens of thousands of individual calculations would rule out "luck" in my mind.

Looking deeper into what Arrhenius had calculated in 1896 and presented to the world in 1908, I came across what I have called in this book the Second Arrhenius Equation and what others have called his "greenhouse law for CO_2":

$$\Delta F = \alpha * \ln (C/C_0)$$

This is not a formulation that Arrhenius himself would have generated. Delta F is the change in radiative forcing, or the strength of the greenhouse effect. C is current carbon dioxide concentration. C_0 is the original concentration. The ln is the natural logarithm function that makes the response nonlinear. Alpha α is a single statistically fit parameter.

Radiative forcing was not a term of use in Arrhenius's time but is what drives changes in temperature. What Arrhenius accomplished was the first quantitative calculations that predicted the impact of changes in carbon dioxide concentration on temperature, both globally and by latitude. In *Worlds in the Making* (1908) his revised estimates for the impact of changes in carbon dioxide described a nonlinear response that was decades ahead of his time. Later formulations have also described this nonlinear response. The Second Arrhenius Equation (and I will keep calling it that in honor of his amazing computational achievement) is one formula that captures the shape of this relationship.

We now have data sets that Arrhenius could only dream of having: several decades of careful measurements of both carbon dioxide in the atmosphere from Mauna Loa, and the globally averaged temperature data covering that same period (see chapter 10). In the classroom setting, wouldn't it be fun and interesting (these are science students after all) to develop a rigorous test of this relationship using these amazing data sets?

This led to the development of a simple spreadsheet-based lab that taught some intuitive basics about statistics as well as testing how well the Second Arrhenius Equation describes the relationship between carbon dioxde and temperature.

The result? The relationship between the measured increase in carbon dioxide in the atmosphere and the measured increase in global annual temperature is surprisingly strong (figure 11.1). The average difference between the predicted temperature and the measured temperature is just a little over 0.08, or eight hundredths of a degree Celsius (compared to a range of more than one hundred hundredths).

With such a good initial relationship, it would be very surprising if there was a second factor that could make that relationship even stronger—but there is one. In Spencer Weart's webpage on the

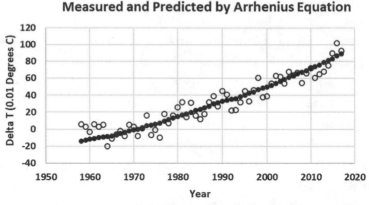

Figure 11.1. Changes in measured global mean temperature (delta T) over time (open circles), as predicted by the Second Arrhenius Equation, with alpha equal to 4.05 (filled circles). The temperatures are in one-hundredths of degrees Celsius.

history of global warming, he mentions the role of El Niño in altering global average temperatures. A strong El Niño means warmer sea surface temperatures in the central Pacific. These warmer temperatures occur across such a huge area that significant warming is translated to the atmosphere and then spread regionally to globally through modifications of global circulation patterns, as discussed in chapter 5. Average global temperatures tend to increase in such a year. Could there be an El Niño signal in those years where the predicted and measured temperatures in figure 11.1 were most different?

There is. Even though the differences are small between the observed values in figure 11.1 (the open circles) and the predicted values (the closed circles forming a nearly continuous line, resulting from the relatively constant increase in carbon dioxide from one year to the next), there is still a relationship between those differences and the

Niño 3.4 index for November, December, and January (NDJ) of the previous year. By including this El Niño index (Niño 3.4, chapter 5, figure 5.3) statistically as a second predictor, the error in the prediction of temperature from carbon dioxide is even lower (0.061, or just over six hundredths, of a degree Celsius).

This simple but strong result then leads to a classroom discussion about correlation versus causation—a warning that just because two measurements plot up well, as in figure 11.1 (they are correlated), that does not prove that one causes the other. There needs to be a physical basis for the causation. In this case, we know very well that increased carbon dioxide should increase global temperatures, and that this relationship should be affected by El Niño. So the scientific basis supports the statistical result.

Can we bring those complex models of climate change into this analysis in some way? Those models use different socioeconomic scenarios that generate changes in atmospheric carbon dioxide and other drivers of climate change, and then predict future changes in temperature. If those predictions also fit this simple relationship, then all the detailed understanding of climate drivers in those models would also support this simpler presentation made possible by the Second Arrhenius Equation. Do they?

In fact they do. While data tables from the Sixth Assessment have not been finalized and released as I write this, carbon dioxide concentrations and projected temperatures for six of the scenarios reported from the IPCC Fourth Assessment are available. Using the same Arrhenius equation with the same value for alpha driven by concentrations of carbon dioxide from the complex model runs yields very accurate predictions of future temperatures as predicted by the models (figure 11.2). This second figure is not the result of a new statistical fit to these future predictions but a test of the accuracy of the statistical fit in figure 11.1 against a new set of data.

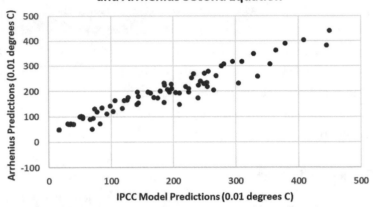

Future Temperatures Predicted by IPCC Models and Arrhenius Second Equation

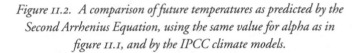

Figure 11.2. A comparison of future temperatures as predicted by the Second Arrhenius Equation, using the same value for alpha as in figure 11.1, and by the IPCC climate models.

Simply put, applying the results of the simple statistical method used for figure 11.1 to predictions of future temperature using only carbon dioxide concentrations generated for those future years matches well against the temperature predictions from the complex models. Thus all the complexity of the detailed models supports the use of the simpler Arrhenius equation as a way to present the impacts of carbon dioxide on climate. The most recent IPCC science report draws a similar conclusion that has strong relevance for managing our carbon and climate future (chapter 12).

Let me point out that the second part of this analysis would not be possible if the complex climate models did not exist. They allow the extension of this simpler index into the future. Said another way, the understanding of climate dynamics as expressed in the complex models makes this simplification possible.

It is not an exaggeration to say that I was stunned by the strength of this relationship. Several of my scientific colleagues have also been very surprised that this relationship is so strong.

No one will claim that carbon dioxide is the only greenhouse gas operating in the atmosphere, although it is the most important. What we appear to have here is a valuable index to the state of the atmosphere that relies on the most basic and clearest understanding of the greenhouse gas phenomenon. It basically says that, as the world works currently, and as projected by the models into the future, the other impacts of human activity on the atmosphere change in concert with, or are consistent with, or are captured by, changes in carbon dioxide.

It is like capturing the state of the Walker circulation and the El Niño system using the single Niño 3.4 measure of sea surface temperature anomalies as an index.

This analysis allows a simple four-step explanation of climate change that I have found to be more convincing than trying to explain how the climate models work:

1. Greenhouse gases have been known for more than a hundred and fifty years to trap longwave radiation (Tyndall in the 1860s).
2. Human activity is increasing the concentration of greenhouse gases in the atmosphere (the Keeling Curve).
3. Measured changes in carbon dioxide are an accurate index to the total impact of all greenhouse gases on global temperature (figure 11.1, with reference back to Arrhenius in 1896).
4. This index in turn accurately predicts increases in global temperatures as projected by the complex global climate models (figure 11.2).

There is a crucial point to be made here about the availability of data. It is the existence of solid, verifiable, long-term data sets such as the Keeling Curve that allows us to see where we have been and where we might be going in terms of climate change.

It can be difficult at times to justify to those agencies that fund research that the continuation of long-term data accumulation is worthwhile. With an emphasis on new and innovative science, the term "monitoring" has taken on a bad aura. The Keeling Curve itself was threatened with termination a number of times, and there was an actual short-term hiatus in 1963. Without that curve, not only would it be impossible to carry out the analysis presented here (a minor concern), we would not really know what the impact of human activity has been on the chemistry of the atmosphere.

The same can be said for other monitoring activities. The National Acid Precipitation Assessment Program (see chapter 9) initiated long-term measurements of precipitation chemistry in the United States. Continuation of those measurements, always under pressure, has been essential to demonstrate the effectiveness of the Clean Air Act Amendments of 1990. Many, many other examples could be given.

Presenting simpler models is one way of engaging audiences on climate change, but if more complex models are to be used, there is an additional step or concept that can help bridge the "trust me" gap. Some call it verification, but I prefer validation. The core idea is this: How accurate have predictions made years to decades ago by similar models been when compared to measurements made after the predictions?

You could say that we have talked about validation in some detail relative to the models used to predict weather when presenting information on the accuracy of those predictions for different numbers of days into the future (chapter 4). Figure 11.2 is also validation of a sort, as the relationship derived in figure 11.1 is applied to a different data set (from the climate models). Essentially, validation means taking a model derived using one set of data and applying it to another, totally independent set.

As another example, the first IPCC report, in 1990, used models to predict the possible range in sea level rise by 2010. A later report compared these predictions with measured values and found actual sea level rise to be at the upper end of the range of predictions. So in this case the models were fairly accurate, if a little on the conservative side.

There have been, over time, a number of other simple models, in addition to the Second Arrhenius Equation used here, that have estimated the impact of greenhouse gases on climate. They do not have the spatial resolution of the big climate models, but to the extent that they yield similar results, they can convey just how long we have known some of the most basic pieces of the climate change puzzle. Like many other parts of the climate story presented here, the predictions of these simple models have not changed a whole lot over the past hundred or more years.

We could start with Tyndall, who offered: "Without water vapor the Earth's surface would be held fast in the iron grip of frost"; but he did not make a quantitative prediction of the impact of different gases on temperature.

Quantitative estimates start with Arrhenius and his prediction that a doubling of carbon dioxide in the atmosphere would lead to a 4°C increase in global average temperature. This relationship between a doubling of this gas and the resulting increase in temperature has come to be called climate sensitivity to carbon dioxide.

Based on the analysis in figure 11.1, Arrhenius was not far off. That average value for alpha in his Second Equation, derived from measured carbon dioxide and temperature data sets, would yield an increase of about 2.8°C for a doubling of carbon dioxide. He was off by just over 1 degree—not bad for 1908!

Both Arrhenius's estimate and the actual data analysis in figure 11.1 include feedbacks, especially the water vapor feedback (the fact that warmer air can hold more water vapor, and water vapor is the strongest greenhouse gas). These would increase the calculated sensitivity when compared with estimates, like some below, that do not include this feedback.

In the 1930s, Guy Callendar used the basic properties of absorption of infrared radiation by carbon dioxide to predict an increase of 2°C for a doubling of carbon dioxide. This estimate was generated in the absence of the water vapor feedback.

That simple, first computer model developed around 1956 by Gilbert Plass estimated carbon dioxide sensitivity at 3.6°C.

In 1967, Syukuro Manabe and Richard Wetherald made the first detailed calculation of the greenhouse effect incorporating convection. They found that, in the absence of unknown feedbacks such as changes in clouds, a doubling of carbon dioxide from the current level would result in approximately a 2°C increase in global temperature.

By 1975, Manabe and Wetherald had developed a three-dimensional global climate model that gave a roughly accurate representation of the current climate. Doubling carbon dioxide in the model's atmosphere resulted in a 2.3°C rise in global temperature. (In 2021, Syukuro Manabe shared the Nobel Prize in Physics for his long-term commitment to modeling the climate of the Earth.)

In 1979 the U.S. National Research Council published a report that concluded: "When it is assumed that the [carbon dioxide] content of the atmosphere is doubled and statistical thermal equilibrium is achieved, the more realistic of the modeling efforts predict a global surface warming of between 2°C and 3.5°C, with greater increases at high latitudes. . . . [W]e have tried but have been unable to find any overlooked or underestimated physical effects that could reduce the currently estimated global warmings due to a doubling of atmospheric CO_2 to negligible proportions or reverse them altogether."

Starting in 1990, the reports of the IPCC have generated the consensus estimates of the impact of carbon dioxide on global temperature. Those numbers have not changed significantly across the thirty-year IPCC history, although the complexity of the models built to generate the numbers has increased dramatically. The IPCC numbers range from 1.5 to 4.5°C, mostly between 2 and 4°C. The consensus figure from the latest science summary is 3°C.

So the best estimates of the increase in global temperature to be expected with a doubling of carbon dioxide, from both simple and complex models, going back more than a hundred years, have been amazingly consistent.

It is not too hard to understand how the carbon dioxide sensitivity alone would not change drastically over time, as the basic properties of this gas in terms of absorption of infrared or heat radiation have been known since the mid-1800s. But why would the relationship in figure 11.1 hold in terms of predicting total atmospheric warming with carbon dioxide alone? Essentially, why is carbon dioxide concentration such a powerful index to all of the changes in the atmosphere caused by all of the components of global change?

To answer that question, we have to take a more detailed look at how changes in the chemistry of the atmosphere are summarized in terms of their effect on surface temperature.

On a hot day, if you are in an air-conditioned space, you are advised to turn off any lights in the space (especially if you still have the older incandescent bulbs as opposed to either compact fluorescent or LED bulbs). Why? Incandescent bulbs use only a small fraction of the electricity consumed to light the filament, with the rest being lost as heat. Those bulbs get hot! LED bulbs in particular convert much more of the energy consumed to light, rather than heat, so they use less energy to create the same amount of light. A six-watt LED bulb can create as much light as a sixty-watt incandescent one and so will be much cooler, consume less energy, and cost much less to run.

So what does this have to do with global warming? Gases and particles in the atmosphere are constantly radiating infrared heat toward the surface. Figure 10.2 captures this process, showing that about two-thirds of the energy reaching the surface of the Earth comes not directly from the sun (163 units) but from infrared radiation given off by components of the atmosphere (340 units).

The units in figure 10.2 and used to measure the increase in heat captured by increased greenhouse gas concentrations are watts per square meter. As greenhouse gases are added to the atmosphere, more of the heat radiation leaving the surface of the Earth is absorbed by those gases, and atmospheric temperature increases. As the temperature of the gases and particles in the atmosphere increases, more heat energy is radiated back to the surface. The impact of this increase in longwave radiation hitting the surface is called radiative forcing— again expressed in watts per square meter.

The impacts of human alterations of the atmosphere can be described in terms of changes in net radiative forcing. In figure 11.3a, gray bars to the right of the center line indicate a warming impact, black bars to the left a cooling impact. The big bar on top is carbon dioxide alone. The next three gray bars include other greenhouse gases like methane and nitrous oxide. The bar for halogenated gases (such as the chlorofluorocarbons discussed in chapter 9) has both positive and negative terms, based on their complex interactions with the atmosphere. Then there are a number of smaller terms, including some that decrease radiative forcing, like cloud formation, dust and aerosols, and changes in albedo due to land use. Total increase in radiative forcing from 1750 to 2015 is estimated to be about 2.3 watts per square meter. That is like adding an extra hundred-watt lightbulb (the incandescent variety) over about every 40 square meters. Sounds small, but multiplied by the surface area of the Earth, that becomes a very big number!

Radiative Forcing Caused by Human Activities Since 1750

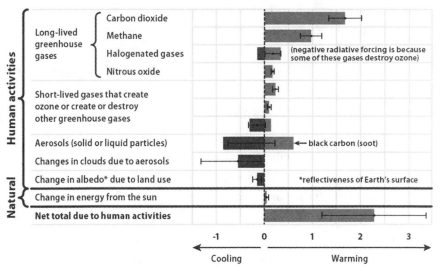

Radiative forcing (watts per square meter)

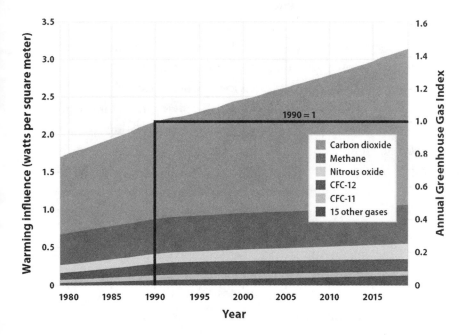

Figure 11.3. (top) The total current radiative forcing for all major drivers of climate change; (bottom) changes in radiative forcing by several different gases from 1980 to 2015.

The concept of carbon dioxide as an index to total radiative forcing is also captured in figure 11.3b. Not only is carbon dioxide the biggest contributor to increased radiative forcing, there is also a fairly constant ratio between the impact of this gas and that of the other major greenhouse factors. The IPCC Sixth Assessment science report projects this similar ratio between carbon dioxide and total radiative forcing through 2100 for the five major climate scenarios, which is why the Second Arrhenius Equation works for modeled futures as well.

This is a long way to get to the concept that the Arrhenius equation works well as an index of climate change. It works because carbon dioxide is the largest part of increased radiative forcing, and there has been, and is predicted to continue to be, something like a constant ratio between the forcing due to carbon dioxide and total forcing. This is the idea of an index. Like Niño 3.4 and El Niño impacts, total carbon dioxide concentration relates well to changes in total radiative forcing caused by all greenhouse gases and other factors.

To summarize, estimates of the impact of carbon dioxide and other greenhouse gases on global temperature have been consistent for more than a hundred years. As the National Research Council said as early as 1979, there do not appear to be any significant missing factors or processes. We can predict the impact of increasing greenhouse gases on atmospheric temperature accurately. That can be done with sophisticated and complex climate models or with simpler statistical relationships supported both by data and the results of those complex models. In my experience, the simpler models are easier to explain and are more convincing.

Knowing that the science of climate change is settled and solid, and can be presented with either simple or complex models, can we look with some scientific confidence at what our climate future might hold, and how predicted changes might affect your daily life?

12

Predictions and Possible Surprises

This guided tour has perhaps convinced you that the science of weather and climate is well established. Chaos will always limit how many days into the future we can predict the detailed machinations of the global energy engine that we call weather. At the same time, an understanding of the long-term changes in the climate system, as driven by increases in greenhouse gas concentrations and constrained by global-scale responses of oceans and ice, allows us to see with some clarity the options we have for shaping our climate future.

Clarity and certainty have increased with each round of the very deliberate IPCC process, and the most recent IPCC science document, released in 2021, provides a definitive set of statements about the state of the science and the state of the climate system. It is a remarkable document. In terms of internal consistency and exhaustive depth, I can think of no other publication on an environmental topic that can approach it. This amazing publication clearly signals that the scientific community has completed the understanding stage of environmental grief and is ready to offer insights on modifying the rate of change or mitigating the consequences.

The major findings from the science report are presented in the forty-two-page "Summary for Policymakers," condensed from the nearly four thousand pages of the full report. I've organized these conclusions around four basic themes: human impacts on climate, momentum and inertia in the climate system, possible tipping points or chances for sudden and unpredictable change, and the certainty that our collective decisions regarding greenhouse gas emissions can alter our climate future.

As your tour guide, let me paraphrase the major findings from the summary, organized into these categories.

Human Impacts on Climate

- Human actions have warmed the atmosphere, ocean, and land. Recent rates of change in the climate system are unprecedented at the scale of hundreds to thousands of years and are creating new weather and climate extremes everywhere.
- Local impacts will be variable, but directional change will be felt in all regions.

Climate Momentum and Inertia

- Many changes caused by greenhouse gas emissions are irreversible, especially those affecting ice sheets, oceans, and global sea level.
- As the climate warms, changes in the climate system will accelerate, and land and ocean sinks for carbon dioxide may become less effective.
- The lowest emission scenarios will require up to twenty years to alter the current trajectory of global temperature increase.

Tipping Points

- The collapse of major ice sheets and sudden changes in ocean circulation are unlikely but not impossible

Modifying Our Climate Future

- The best estimate of climate sensitivity to a doubling of carbon dioxide is 3°C.
- Limiting global warming to 1.5°C or 2°C will require deep reductions in greenhouse gas emissions.
- There is a direct relationship between total cumulative carbon dioxide emissions and increases in global average temperature.

*

The first category is the primary conclusion from the IPCC process, and one that has been apparent for decades: we are changing the global climate system. The Earth is warming, and while the impacts of this will vary regionally, everyone everywhere will be affected.

Global trends will include higher maximum temperatures, fewer days with frost, increased precipitation, and especially an increase in major precipitation events. Soil moisture (drought) is more complex. More rain would tend to diminish drought, while longer growing seasons and higher temperatures might enhance drought. In either case, more intense precipitation events should mean increases in flood events as well. In addition to these global-scale patterns, the full science report presents detailed maps projecting impacts for every region.

The full IPCC science report from 2021 and the complete Sixth Assessment released in 2022 present examples of many of the tragic consequences that could result from projected changes in the climate system. The details of these scenarios of human suffering are beyond the scope of this book but provide the critical context for any discussion of climate dynamics.

In many ways, major impacts of climate change are already being felt. In the United States, parts of Miami already experience periodic flooding even in the absence of storms. Under the highest emissions scenarios, much of the southern third of Florida could be below sea level by the end of the century. The question of reinvesting in and re-populating flood-damaged areas, such as the parts of New Orleans destroyed by Hurricane Katrina, has already become a controversy that pits a sense of place and home, as well as wealth, race, and culture, against the recurring costs of insurance and reconstruction. The same questions can be asked about areas in the western United States recently ravaged by major wildfires that included tragic losses of life and property.

*

One of the most sobering revelations for me in putting together this paraphrased summary is captured in the second category: the climate system has developed a momentum that will be impossible to reverse, and the climate response to even major reductions in emissions will not be detectable for two decades. From previous chapters, you won't be surprised that this momentum resides in the oceans and in ice, as these are the most slowly changing components of the climate system. In chapter 10, these two were credited with delaying the full impacts of greenhouse gas emissions, through inertia, if you will, in their response to warming.

We should recognize and understand that the long-term momentum now in place is the flip side of the reprieve that oceans and ice have given us through this short-term climate inertia.

Just how much this inertia has prevented more rapid changes in perhaps the most challenging impact of climate change, sea level rise, can be suggested by looking back in geologic time to previous eras with average global temperatures similar to those predicted for the next century.

IPCC projections for increases in global mean temperature by 2100 range from about 2°C at lower emission rates to as much as 5°C at the highest levels. When was the last time that the world was 2 or

5 degrees warmer than at present, and what were sea levels at those times? For 2 degrees warmer, that was about five million years ago, and sea levels were about 20 meters higher. For 5 degrees, the answers are forty to fifty million years ago and about 70 meters higher, in large part because Antarctica was ice free.

No one says we are looking at anything approaching this level of change on a human time scale, but it is the inertia present in the slowly melting ice caps in Greenland and Antarctica that are preventing it. The IPCC science report suggests that sea levels might rise by as much as a meter by 2100. The difference between a projected rise of 1 meter in sea level resulting from a 2 degree increase in temperature and the 20 meter rise that occurred the last time the Earth was two degrees warmer tells us just how important that ice-based inertia is in the climate system. At current rates of ice loss in Antarctica, it may take more than a hundred and fifty thousand years for that ice cap to disappear completely.

For perspective, though, remember that all the recent reports of disruptions related to sea level rise have been in response to measured increases of only 8–12 inches (20–30 cm).

The apparently irreversible momentum in the climate system suggests that we should extend the time line of our climate projections. Most climate change scenarios and emissions goals look to 2030 or 2050, with a few extending out to 2100. This makes sense in telling the story to those of us who are in charge of the world at this point. The twenty-second century may seem a long way off, but children born today have a good chance of seeing 2100, and their children and grandchildren will see 2150 and beyond.

This latest IPCC science report does go past 2100 for sea level predictions, projecting to 2300. At that point, predicted sea level rise centers around 1.5 meters for the lowest emissions scenario and around 4.5 meters for the highest scenario. There are large uncertainties around those estimates, including the possible if unlikely enhanced melting of the ice caps that could push sea level rise to as much as 15 meters by 2300.

There are no scenarios that predict a decrease in sea level.

*

The Sixth Assessment science report generally depicts a continuous and progressive change toward a very different climate, with important consequences. Are there places in the climate system that we might want to monitor most closely, those where change might become discontinuous or accelerated through the kinds of positive feedbacks discussed in previous chapters? Are there potential tipping points?

The Atlantic meridional overturning circulation (AMOC) took center stage in both chapters 5 and 6. In chapter 5, the Gulf Stream was described as crucial to the historically consistent and anomalously warm temperatures experienced in northern Europe. In chapter 6 the ice core record presented by Richard Alley in *The Two-Mile Time Machine* described abrupt changes in climate over Greenland, at the northern end of the Gulf Stream, which could be associated with changes in the rate of flow of the Gulf Stream, and a slowing of AMOC and of ocean circulation globally. Could this happen or is it happening now?

Direct measurements of flow for the Gulf Stream are available for about fifteen years, not long enough to separate directional change from periodic cycles. NOAA, however, has derived a long-term data set on changes in ocean temperature at the northern end of AMOC, near Greenland, that could reflect rates of flow. That record shows some consistent cooling over the past decade.

Because of the vast size of this major vector for transporting tropical heat northward, we might expect it to change slowly. The science report from the Sixth Assessment makes a longer-term prediction of a continuous slowing of the AMOC circulation, extending out to 2300. That projection includes a greater reduction in AMOC and the flow of the Gulf Stream by 2100 than any presented in that longer NOAA data set going back to 1860.

A rapid and major shift in the global ocean circulation system is one of the abrupt changes considered as a risk factor in the IPCC science report. Disruption of AMOC and the Gulf Stream could drive that change. There is some solid science behind the concept of rapid shifts in the Gulf Stream.

The potential for accelerated melting of the Greenland and Antarctic ice caps has already been described, but another potentially important feedback or tipping point involves a very different kind of frozen water.

The extent of permanently frozen soils (permafrost) was described in chapter 10. The total amount of carbon stored in permafrost is estimated to be about twice the amount currently in the atmosphere. Rapid loss of permafrost in the Arctic has been documented.

The Sixth Assessment predicts that much of this permafrost area will no longer be permafrost by the end of the century, but again the range of predictions varies widely across the different scenarios. The fate of the carbon in the organic matter in these newly activated soils is important. Will it remain stable—like long-term storage of humus in soils throughout the world? Will most of it decompose, releasing the carbon it contains? If it does decompose, will that carbon be released as carbon dioxide or as methane (a much more powerful greenhouse gas)? Here is another Earth system process we need to monitor closely.

An additional area of climate uncertainty may well involve the oscillations presented in chapter 5 that shake our figurative climate parachute. In the gray area between the short-term modeling of weather and the longer-term projections for climate, our relative lack of understanding and predictive capacity about these oscillations stands out as having the potential to surprise us. We need to know more about these movers and shakers in the weather/climate interface.

So there are components of the climate system that bear close watching. This only underscores the importance of the global network of private and public organizations, and the scientists and technicians who work with them, that provide the continuing measurements that make up the key data sets related to climate change. These accumulating data sets are absolutely essential for understanding our climate trajectory. This large and diverse community deserves an inclusive thank-you for all the diligence and hard work brought to this critical field of science.

I will venture to say that any and all of the uncertainties described here lean in the direction of more rapid rather than less rapid warming. The scientific community continues, as with the IPCC process, to be conservative about predicting rapid change.

And there is one additional crucial variable that is embedded in the deeper layers of the future scenarios used in the IPCC process to predict our climate future but is otherwise missing from much of the climate change discussion: population growth.

There is very little in the IPCC science reports about the number of people the Earth can sustain, and sustain at a reasonable standard of living. In chapter 9 I presented Malthusian and cornucopian views of population growth and our ability to invent solutions to sustain ever-increasing numbers. There is a clear connection to climate change as well.

The proximate cause of global warming is the accumulation of greenhouse gases in the atmosphere. One step toward describing the ultimate cause of this warming is the link between concentrations and emissions, driven primarily by the burning of fossil fuels, but also by agriculture. What is the ultimate cause of those emissions? Surely it is global economic activity and the efficiency with which we use fossil fuels and agricultural resources like fertilizers to drive that activity. But there is one more step back to a crucial ultimate cause of emissions.

A simple equation for total emissions (in the spirit of the Second Arrhenius Equation) would be:

Total emissions = N multiplied by $\$/N$ multiplied by GHG/$\$$.

Or the number of people in the world (N) times the level of economic activity per person ($\$/N$) times the GHG efficiency of that economic activity (amount of greenhouse gases emitted per unit of economic activity—GHG/$\$$, or Euro, or Yuan).

Much effort goes into maximizing economic well-being per person ($\$/N$), and rightly so. Much of the current discussion of minimizing climate change focuses on reducing emissions per unit of economic activity (GHG/$\$$, see *Avoiding a Climate Disaster*, by Bill Gates). Very little is said about the desirability or feasibility of limiting population growth (N). Most discussions start with "In order to sustain a population that will grow to xx billion by 20xx we need to . . ."

In fact, one can find articles focused on the problems resulting from aging populations in Europe, Japan, China, and other regions where population decline is imminent. Some would link future economic growth to population growth. Those presentations tend to ignore both the use of resources needed to sustain more of us and the environmental impacts, including greenhouse gas emissions, resulting from that usage.

As with the Second Arrhenius Equation, the equation above for total emissions is testable. Presenting this equation to students in an introductory environmental science class led to a challenge to do just that. Do the data support the importance of population change in greenhouse gas emissions?

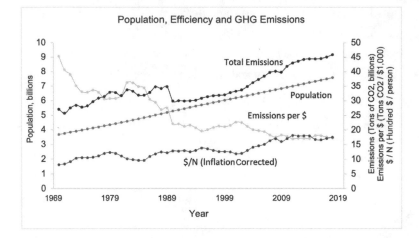

Figure 12.1. Human populations (in billions, left axis), economic activity per capita (in $100s, right axis, corrected for inflation), carbon dioxide emissions per unit of economic activity (in tons per $1,000, right axis), and total carbon dioxide emissions (in billions of tons, right axis).

Figure 12.1 charts the interactions of the variables in this equation globally since 1970, based on data from the World Bank. Even though emissions per unit of economic activity have declined by about 60 percent, a near doubling of population and some growth in real income per person (adjusted for inflation) has led to an increase in total emissions of more than 60 percent.

The factors controlling, encouraging, or limiting human numbers are far beyond the scope of this book, and I realize that this is a tremendously oversimplified view of a very complex issue. It is included here just to suggest that providing food, housing, and a decent standard of living for a population that may increase by another 50 percent by 2100, if that is what happens (and there I go with the same approach criticized above), will counteract whatever is done to increase the carbon efficiency of the economic system.

Perhaps the food system should be of special concern. Current estimates suggest that 10 percent of total greenhouse gas emissions in the United States and as much as 33 percent of global emissions come from agriculture, driven largely by higher emissions of methane and nitrous oxide relative to other sectors of the economy.

It is well documented that global eating patterns turn more toward carnivory as standards of living increase, and that the production of animal-based foods, especially beef, generates several times more greenhouse gas emissions per food calorie than grains. If we are effective at reducing greenhouse gas emissions from fossil fuels, then agriculture, driven by increases in the human population and increased carnivory, could loom as the most difficult next step.

It could also break the relationship in figure 11.1 between carbon dioxide concentrations in the atmosphere and global average temperature. The Second Arrhenius Equation could lose its relevance as methane and nitrous oxide become the dominant greenhouse gases instead of carbon dioxide.

In 2013 I had a chance to speak with the Board of Directors of the Environmental Defense Fund. The topic was supposed to be a general discussion of biogeochemical cycles, but this well-informed group focused very quickly on agriculture. There was serious discussion that the global food system was the next important area for realizing environmental improvement. The current generation of students definitely sees the connection, and the level of interest in sustainable agricultural methods, along with the business opportunities in a more localized food system, runs high among them.

*

The certainty of climate change, the momentum present in the climate system, and the potential for tipping points to accelerate that change all present a somber picture of our climate future. Are there ways to alter that future?

There are many. Decarbonizing the economy is a hot topic. The variety of pathways proposed to bring us to a low- or zero-carbon future is astounding and leads us into the dynamic and complex areas of energy and food policy, taking us again beyond the scope of this book.

But there is a summary relationship derived from the IPCC process that can quantify the potential impact of any carbon reduction strategy. That relationship also brings us back to the Second Arrhenius Equation, providing a historical bookend to our guided tour.

The best estimate of climate sensitivity to a doubling of carbon dioxide included in the set of summary conclusions above (3°C) is right in the middle of estimates going back more than a hundred years, as presented in chapter 11. And the conclusion that there is a direct relationship between cumulative carbon dioxide emissions and increases in global temperature harkens back to figures 11.1 and 11.2, and to the monumental calculations of Svante Arrhenius in the late 1800s.

Figure SPM.10 in the "Summary for Policymakers" from the Sixth Assessment science report presents graphically this direct relationship between increased globally averaged temperature and cumulative carbon dioxide emissions. A single relationship applies both to measurements made up to the present time and to projections into the future by the climate models. It presents alternate pathways to our future in a way that both Arrhenius and Occam might approve. To the extent that we reduce carbon dioxide emissions, there will be a direct reduction in warming.

This figure in the summary also specifies the degree of reductions in emissions required to reach the goal of limiting global temperature increases to 1.5°C or 2.0°C, benchmarks often cited as crucial for limiting the negative impacts of warming. For perspective, other graphics from the summary specify that keeping within either of those target temperatures through 2100 will require net negative carbon emissions beginning sometime between 2050 and 2100. The notion of net negative carbon emissions assumes that "sinks" for carbon dioxide, like forest growth or technologies for actually removing greenhouse gases from the atmosphere, become greater than total emissions.

And how are we doing so far? One recent report projects that current policies put in place by countries around the world would only bring emissions in 2100 back to what they were in 1990, leading to a 2.5–2.9°C increase in temperature. Recent pledges to reduce emissions, if realized, would reduce emissions to about one-third below the 1990 baseline and increase temperatures by about 2.1°C by 2100.

Both the most recent IPCC analyses and the simpler approach we have developed inspired by Svante Arrhenius (figures 11.1 and 11.2) tell us the same thing: reducing carbon dioxide emissions will slow the rate of warming (with the caveat about greenhouse gases from agriculture as already described). This would not reverse the momentum in the climate system, but it would slow the rate of warming and of sea level rise, allowing more time for adaptation.

The biggest unknown in predicting our climate future is not the science, it is in what the global community decides to do about greenhouse gas emissions.

*

Every climate response presented in this chapter describes a range of futures that depend on what the world does about greenhouse gases. There will be no shortage of individual solutions proposed. Both the "Summary for Policymakers" from the IPCC assessments and Bill Gates's *How to Avoid a Climate Disaster* offer pathways to a different future. John Houghton's *Global Warming* discusses some of these options in detail, whereas Kerry Emanuel's *What We Know about Climate Change* offers a briefer summary.

I thought it likely that the release of the complete Sixth IPCC Assessment in February 2022, with its implications for global economic, energy, food, and population systems, would reignite all of the controversies and arguments typical of the anger and denial stages of environmental grief (chapter 9). Events that February rightly put war-torn Ukraine rather than climate change at the top of the world's agenda. Still, there will likely come a time soon when the persistent impacts of the changing climate system bring climate back into public focus.

Recent history suggests that we as a society appear to prefer to respond to a climate change crisis than to prevent it. We should understand that either prevention of climate catastrophes or recovery from them will be expensive. I remember distinctly hearing a Dutch hydrologist say, on a news broadcast following Hurricane Katrina, that we could have prevented the tragedy in New Orleans. He was visibly in tears as he said it, and the Dutch know a thing or two about keeping the ocean at bay. Protection against rising seas can be accomplished in some places, but it won't be cheap. Neither will rebuilding following disasters.

The first step in addressing this complex set of issues is to agree that climate change is happening. Any discussion about how to prepare for climate change is more likely to succeed if there is some basic agreement about the underlying science. I hope that this book, and others like it, can contribute in some small way to an evaluation of the inevitable policy and political arguments ahead of us based on a deeper and more general scientific understanding of the climate system.

I end this chapter, and our guided tour of weather and climate, where it began: with a return to the historical perspective.

It should be clear that the scientific community is well beyond the understanding stage of environmental grief and hoping to help with solutions. This understanding, while newly sharpened, is not newly attained. Tyndall told us of the greenhouse gas property of carbon dioxide and other gases in the 1860s. Arrhenius provided a first and rather accurate estimate of the impact of increased carbon dioxide on global temperature, and its spatial distribution, in 1896 and presented his conclusions to the public in 1908. Charles Keeling showed us the integrated impact of human activity on carbon dioxide in the atmosphere starting in the mid-1950s, and James Hansen described what this meant for our climate future in 1988. The clear relationship between carbon dioxide emissions and global temperature presented in the most recent IPCC science document, drawn from thirty years of solid climate science research and synthesis, shows how the path ahead can be modified.

Potential surprises described here due to oscillations and tipping points present interesting intellectual challenges, but they are more likely to accelerate rather than moderate our current trajectory. All of the possible feedbacks described involving ice and oceans would suggest that, if anything, those predicted trajectories are conservative.

So our tour is ended, and the science is clear. We should put what we now know about weather, climate, and climate change to work for us to alter our climate trajectory and prepare for our climate future.

EPILOGUE
Making It Local

One undeniable truth about global change is that it will affect each area or region differently. In the United States, Florida may be concerned about flooding. Arizona might worry about excessive heat. Southern Californians might want to understand how a changing climate system could impact El Niño and the recurring relief that it has brought to a frequently drought-stricken region—relief from drought at the expense of major storm damage.

Beyond the global impacts of climate change, it might be local changes that affect where we live that bring home to most of us the impact of global climate change. One of the most frequent questions I hear is something like this: "But what does this mean for me here where I live?"

So in this epilogue let's look at one effort to downscale the predicted impacts of global climate change to a particular region. I hope you will excuse me for focusing on the place where I live: New Hampshire. You should be able to find similar reports for the place where you live.

How are fine-scale local and regional impact projections made? It's a two-step process. The first step is to take the global climate predictions, at the relatively coarse scale they are generated, and add finer-scale specificity to generate a more realistic local map. The simplest statistical way to do this is to relate the average value predicted for that larger area to the finer-scale measurement of temperature and precipitation from all the weather stations within it. This allows an extension of the single number to the site-specific kind of weather patterns we experience. Yes, it will be cooler at higher elevations or near a coastline or large lake (in summer anyway).

The next step is to translate the resulting maps into climate and weather changes that affect activities important to the people in that area. For my example, central air conditioning has not traditionally been needed or generally used in homes in New Hampshire. The ski season has been long and profitable, and not all ski areas required expensive snowmaking equipment. This has changed in the thirty-five years I have lived here. On the other hand, small-scale agriculture is thriving as local farmers' markets grow and most restaurants, not just the pricey ones, increase reliance on local food sources.

What does the future look like here in New Hampshire for these sectors, as they are called? In a recent report, the Climate Solutions New England project produced some insights into impacts on these major regional activities (figure E.1). Again, I don't think any of these will surprise you now, but they may have greater personal impact. Values are shown for high and low emissions scenarios from a previous IPCC report. I have extracted just the values averaged for all of southern New Hampshire, but these are similar to, and available for, individual cities, and for northern New Hampshire as well.

As average temperatures rise, the number of days with maximum temperatures above 90°F (32°C; that is hot for northern New England!) will increase disproportionately—from fewer than ten to as many as fifty days per year. In chapter 5 I made the distinction between "this has been an unusually hot summer" and "time to put in air conditoning because summers are getting hotter." The top left panel in figure E.1 supports the latter.

On the other hand, a warmer world will mean a longer growing season (up to 30 percent longer, top right) for those farm-to-table

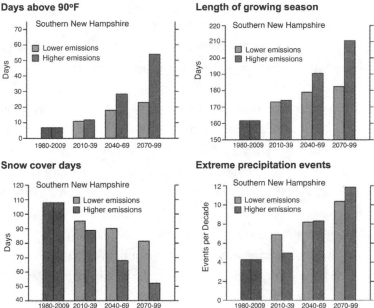

Figure E.1. Predictions of change in some climate factors that affect life in southern New Hampshire. From top left to bottom right: the number of days with a maximum temperature over 90°F (32°C); the length of the growing season; the number of days with snow cover; and the number of extreme precipitation events (more than 4 inches or 10 cm of water-equivalent precipitation in forty-eight hours).

growers in the region. My home institution has an active research program on growing kiwifruit in the state, as well as many projects using high-tunnel greenhouses to extend the growing season even further. New Hampshire is more than 80 percent forested land, and yet these innovative approaches to agriculture are vibrant and growing.

If, however, you are a ski mountain operator, or a snowmobiler, the news is not so good (bottom left). The number of days of natural snow cover is predicted to decline substantially—from about 110 days to as few as fifty days per year. All the major ski areas in the state now depend on recurring snowmaking and grooming operations, and that will continue and intensify into the predicted future. Making and grooming snow is already one of the biggest cost factors for these enterprises.

A final prediction here (bottom right) is an increase in the number of intense precipitation events, defined as more than 4 inches (10 cm) of rain or equivalent water content of snow over a forty-eight-hour period. This may be good news for those who run snow plowing operations, assuming some of that intense precipitation still arrives as snow, but in a region of shallow soils and extreme topography, this also suggests more flooding events. We still live with the damage caused by Hurricane Irene in 2011; and the impact of the Hurricane of 1938, the last to hit the region before radar, and therefore hit without warning, is the stuff of legend.

There are other predictions in this report, including the timing of ice-on and ice-off on major lakes (a shorter outdoor hockey and ice fishing season), but an earlier publication from the Climate Solutions New England group included a graphic that I think captures well the fully integrated effect of all of the predicted changes (figure E.2). Depending on the scenario and the time line used, patterns of weather and climate in my home state will increasingly match patterns now seen in much more southern regions.

I have on occasion said to my young undergraduate students, only somewhat facetiously, not to worry about planning on a move south for a warmer retirement, that warmer weather will come to you.

But these average changes in climate indicators may not be the most important impact if you want to live near the coast. Many major northeastern U.S. cities, from New York to Providence to Boston and on to Portsmouth, New Hampshire, and Portland, Maine, were founded on access to harbors and global trade in the age of sail. Each

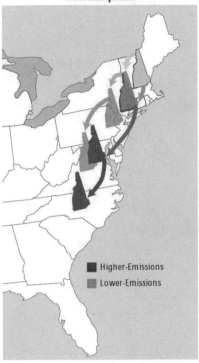

Figure E.2. This figure illustrates how New Hampshire's climate could shift under different future emissions and warming scenarios to resemble the climates of states that today are considerably warmer than New Hampshire.

will face rising sea levels and threats of flooding. Inexorable sea level rise is perhaps the most serious issue for every city near sea level worldwide.

No matter how the information is presented, we will all be better prepared for our climate future by being aware of the ways in which that future could affect the pattern and quality of our daily lives.

Bibliographic Essay

The references below are not rich in primary articles from technical journals. Scientists writing for technical journals are required to use a form of shorthand or jargon that communicates precisely and succinctly but tends to make the results undecipherable for those outside the discipline.

Most of the sources mentioned here are books or websites. Updating a website is one of the fastest ways to communicate recent results in science. While social media platforms may be even faster, I do not consider them accurate or useful and do not use them. The quality of reporting on websites can be variable as well, but the good ones provide interesting and informative presentations understandable by general audiences.

The problem with websites is that the links can be easily and quickly broken. In a printed book, listing URLs for particular websites may not be helpful, as many will change shortly. While some specific URLs are included here, in many cases I have provided a descriptive phrase that would lead you to the right page, or something close to it. Exceptions to this approach include citations for figures, as these need to be precise as of this writing.

For this book, the most useful web pages have been those of organizations like NOAA (National Oceanic and Atmospheric Administration—home to the U.S. Weather Service), NASA (which supports satellite-based methods, and also a lot of global change research), and other U.S. government agencies. Also featured are the reports of the IPCC (Intergovernmental Panel on Climate Change) and other international scientific institutions, universities, and publishers.

I have made extensive use as well of pages in Wikipedia, a unique and uniquely valuable source for basic scientific information on topics that are not hugely and hotly controversial. Wikipedia is essentially an information commons, crowdsourced if you like, and driven by the expertise of volunteer editors and contributors. As of this writing, according to the Wikipedia page on Wikipedia, there are more than five thousand volunteers who enter more than a hundred edits each per month. Perhaps it is the vigilance of these editors that keeps Wikipedia from falling to the level of disinformation prevalent on social media sites.

The activity of these contributors and editors leads to the other unique aspect of Wikipedia. While content does not change often on pages dealing with settled topics, change will happen on topics where scientific consensus is evolving (like the Snowball Earth discussion in chapter 7). It is almost like watching the science evolve in real time—watching the experts hash out new ideas online.

The breadth and quality of content on Wikipedia pages, defended if you like by those volunteer editors, are reflected in the fact that Wikipedia (again, according to its own Wikipedia page) is the thirteenth most visited site on the Web, with 1.7 billion unique visitors per month. That represents more than 20 percent of all the people in the world.

For another example of the positive power of crowdsourcing, look for the article "When Pixels Collide" on DigitalCulturist.com. It describes Reddit's experiment in crowdsourcing a work of art. Reddit opened a blank space (r/place) to its anonymous users and allowed all of them to place a colored pixel at any location, wait a few minutes, and then place another. There was chat initially that the space would be covered with the kind of negative trash that occupies so much of social media, but, wonder of wonders, what emerged was coordinated efforts by groups to put up symbols and emblems of their countries or organizations, favorite works of art (the *Mona Lisa* was a prominent image), or favorite cartoons!

The space evolved very rapidly, and because of the time delay between placement by any one person, defining a part of the space took coordinated efforts by collectives working together. As anyone could write over what you had done in any pixel, it also took coordinated defense and repair of the space your group had claimed.

I show this in my classes to demonstrate the power of positive large-group collective action against the mostly small-group negativities that claim way too much of our online consciousness and way too much coverage in the mainstream media. Call it online democracy if you like.

On the Bookshelf

Students in my classes are exposed to my addiction to popular science books. I will bring one or more to most classes, hoping to spark an interest in those bright young minds. Many of these volumes have been used to put together the stories I present to my students, and also to you here. In my introductory course I start many classes by saying "You could write a book on this topic—and some have." It is a way of saying that we will be hitting the highlights and not getting into the weeds (to put a positive spin on it). An opening slide will show and list some of the books used for that class and the suggestion to read them. Sometimes that happens and I hear back from students about it. Sometimes there is a student in the class who says—"Yes, I read that last summer!" Bingo!

The same is true here. There could be, and have been, entire books written on each chapter topic. I have drawn freely on those books and give full credit when doing so. What I hope to contribute with this book is linkage and connection among those topics presenting a complete, integrated story of weather, climate, and climate change—a guided tour through a museum of science on those three interlinked subjects.

Here is a list of the books that are referenced in the chapters, and some others that have contributed to the framing of this book. These are alphabetical by author.

Ahrens, C. Donald, and Robert Henson. 2019. *Meteorology Today.* 12th ed. Boston: Cengage.

Allen, Richard B. 2014. *The Two-Mile Time Machine: Ice Cores, Abrupt Climate Change and Our Future.* Princeton: Princeton University Press.

Alvarez, Walter. 2015. *T. rex and the Crater of Doom.* Princeton: Princeton University Press.

Arrhenius, Svante. 1908. *Worlds in the Making.* Translated by H. Borns. New York: Harper and Brothers.

Blum, Andrew. 2019. *The Weather Machine: A Journey Inside the Forecast.* New York: HarperCollins.

Bowen, Mark. 2005. *Thin Ice*. New York: Henry Holt.

Bryson, Bill. 2003. *A Short History of Nearly Everything*. New York: Doubleday.

Carson, Rachel. 1951. *The Sea around Us*. Oxford: Oxford University Press.

———. 1961. *Silent Spring*. Boston: Houghton Mifflin.

Crawford, Elisabeth. 1996. *Arrhenius—From Ionic Theory to the Greenhouse Effect*. Canton, Mass.: Watson.

Diamond, Jared. 1997. *Guns, Germs, and Steel: The Fates of Human Societies*. New York: W. W. Norton.

———. 2011. *Collapse: How Societies Choose to Fail or Succeed*. London: Viking Penguin.

Emanuel, Kerry. 2018. *What We Know about Climate Change*. Cambridge, Mass.: MIT Press.

Fagan, Brian. 2000. *The Little Ice Age*. New York: Basic Books.

———. 2009. *Floods, Famines and Emperors: El Niño and the Fate of Civilizations*. New York: Basic Books.

Fleming, James Rodger. 1998. *Historical Perspectives on Climate Change*. Cambridge, Mass.: Harvard University Press.

Goldsmith, Barbara. 2005. *Obsessive Genius: The Inner World of Marie Curie*. New York: Norton.

Gore, Al. 2006. *An Inconvenient Truth: The Crisis of Global Warming*. New York: Rodale.

Gould, Stephen J. 1987. *Time's Arrow, Time's Cycle: Myth and Metaphor in the Discovery of Geological Time*. Cambridge, Mass.: Harvard University Press.

Hallam, A. 1985. *Great Geological Controversies*. Oxford: Oxford Science Publications.

Holmes, Richard, and Gene Likens. 2016. *Hubbard Brook: The Story of a Forest Ecosystem*. New Haven: Yale University Press.

Houghton, John. 2015. *Global Warming: The Complete Briefing*. 5th ed. Cambridge: Cambridge University Press.

Imbrie, John, and Katherine Palmer Imbrie. 1979. *Ice Ages: Solving the Mystery*. Cambridge, Mass.: Harvard University Press.

Kean, Sam. 2017. *Caesar's Last Breath*. New York: Little, Brown.

Kimmerer, Robin Wall. 2015. *Braiding Sweetgrass: Indigenous Wisdom, Scientific Knowledge and the Teaching of Plants*. Minneapolis: Milkweed.

Kolbert, Elizabeth. 2015. *Field Notes to a Catastrophe: Man, Nature and Climate Change*. New York: Bloomsbury.

————. 2022. *Under a White Sky: The Nature of the Future*. New York: Crown.

Kuhn, Thomas. 1970. *The Structure of Scientific Revolutions*. 2nd ed. Chicago: University of Chicago Press.

Lovelock, James. 1979. *Gaia: A New Look at Life on Earth*. Oxford: Oxford University Press.

McFadden, JohnJoe. 2021. *Life Is Simple: How Occam's Razor Set Science Free and Shapes the Universe*. New York: Basic Books.

Monmonier, Mark. 1999. *Air Apparent: How Meteorologists Learned to Map, Predict and Dramatize Weather*. Chicago: University of Chicago Press.

Oreskes, Naomi, and Erik Conway. 2010. *Merchants of Doubt*. New York: Bloomsbury.

Rodhe, Henning, and Robert Charlson (eds.). 1998. *The Legacy of Svante Arrhenius: Understanding the Greenhouse Effect*. Stockholm: Royal Swedish Academy of Sciences.

Sagan, Carl. 1980. *Cosmos*. New York: Random House.

Sobel, Dava. 2007. *Longitude: The True Story of a Lone Genius Who Solved the Greatest Scientific Problem of His Time*. New York: Bloomsbury.

Strevens, Michael. 2020. *The Knowledge Machine: How Irrationality Created Modern Science*. New York: Liveright.

Thomas, Lewis. 1974. *The Lives of a Cell*. New York: Penguin Random House.

————. 1983. *The Youngest Science*. New York: Penguin Random House.

Turekian, Karl. 1996. *Global Environmental Change: Past, Present and Future*. New York: Prentice Hall.

Tyson, Neil deGrasse. 2017. *Astrophysics for People in a Hurry*. New York: W. W. Norton.

Weart, Spencer. 2003. *The Discovery of Global Warming*. Cambridge, Mass.: Harvard University Press.

On the Web

Again, there are very few URLs provided here, as those links could be broken at any time. Instead, organizations, page titles and/or major search terms are listed where appropriate for each chapter. Let's hope NOAA, as one example, will always have an El Niño page!

Sources by Chapter

CHAPTER 1. WAS SVANTE RIGHT?

An excellent presentation of the life and work of Svante Arrhenius is contained in Elizabeth Crawford's book *Arrhenius—From Ionic Theory to the Greenhouse Effect* and her chapter in the volume edited by Henning Rodhe and Robert Charlson entitled *The Legacy of Svante Arrhenius: Understanding the Greenhouse Effect*. Most of the information on the life of Arrhenius is abstracted from these sources.

I find no pages for Stockholm Högskola—these have been subsumed under Stockholm University, but the original name is included in the Wikipedia page on the history of the university. There is currently an institution called University College Stockholm (or Enskilda Högskolan Stockholm), which focuses on teaching theology and human rights.

The Wikipedia page for Svante Arrhenius includes the pioneering statements about carbon dioxide and climate listed near the end of this chapter.

There are Wikipedia entries for Vilhelm Bjerknes, Arvid Högbom, and Nils Ekholm. Interestingly, all refer to Stockholm University rather than Högskola. The only reference to the Stockholm Physics Society I could find was a short piece through the *Encyclopedia Britannica* site. I hope the presentation here in chapter 1 will refocus some attention on this unique intellectual institution.

The quotations of Arrhenius's understanding of the influence of carbon dioxide on global temperatures come from his *Worlds in the Making*, pp. 46–63.

CHAPTER 2. OCCAM'S RAZOR AND THE CASE FOR
SIMPLER EXPLANATIONS

Information on William of Occam, Occam's razor, and the philosophical discussion between razor and anti-razor can be found on Wikipedia, with lots of references if you want to go deeper into the topic. I could find no reason why William's name appears sometimes as Occam (the Latin form) and other times as Ockham. I used the most commonly found form. The recent book by JohnJoe McFadden is an excellent treatise on the influence of the search for simpler solutions on the development of our understanding of physics from the cosmos to subatomic particles.

In addition to the Wikipedia page on Vilhelm Bjerknes, there are pages on the Primitive Equations that also link to pages on Atmospheric Flow and Atmospheric Models. There is also a page for Lewis Fry Richardson.

The role of the Stockholm Physics Society in the development of the Primitive Equations by Vilhelm Bjerknes is in Crawford's biography of Arrhenius.

A search for Sagan Planet Walk should take you to a map that shows how the planet plaques are distributed.

As cited in the text, Andrew Blum's *Weather Machine* captures the challenge of accessing the data and running the models to produce short-term weather forecasts.

In *Air Apparent*, Mark Monmonier presents a detailed description of how Richardson went about simplifying the scope and detail of his calculations in order to be able to complete the calculations. Monmonier also captures the rigidity and inertia in the U.S. weather prediction community in terms of resisting the dynamical approach of Bjerknes and the Bergen School, opting instead for qualitative and intuitive interpretations of weather station data when making forecasts.

The Richardson book is: Lewis F. Richardson, *Weather Prediction by Numerical Processes* (Cambridge: Cambridge University Press, 1922).

If you are interested in the editorial, here is the citation: J. D. Aber, "Why don't we believe the models?" *Bulletin of the Ecological Society of America* 78 (1997): 232–33.

There is much more on the European model in chapter 4. Or you can go to the Wikipedia page on the European Centre for Medium-Range Weather Forecasts.

CHAPTER 3. THE BASICS OF WEATHER
AND CLIMATE—IT'S SIMPLE!

As this chapter presents a basic description of the forces driving weather and climate, there are lots of textbooks and Wikipedia pages that cover the basic phenomena. I hope my summary will both save you from having to try to read through all of them to develop the kind of "simple" explanation presented in this chapter, or possibly inspire you to try to find out more about each topic by diving into those pages.

The Wikipedia page on Atmospheric Circulation has a good description and graphical depiction of the Hadley, Ferrel, and Polar cells (although the Ferrel cells are termed "mid-latitude cells" in some of the graphics). Figure 3.2 is from this page.

The Wikipedia page on the Coriolis Effect is a little too technical for easy consumption, but a search for videos on this phenomenon will get you some good demonstrations. In this case, a good video is worth at least a thousand words. You can add "merry go round" to the search to get an example of what is described in the text.

The Wikipedia page entitled Jet Stream is clear and links the process leading to jet streams to the existence of the Hadley, Ferrel, and Polar cells. Figure 3.3 is from this page.

To avoid repetition, I will just say that Wikipedia also has good pages on Rain Shadow, Temperate Rainforest, Palouse, Sea Breeze, Monsoon, and Dew Point and Humidity, with both absolute and relative humidity covered in the same page.

See also pages on Tornado and Thunderstorm, and even Nor'easter! That last page currently includes quite a bit of history and a list of famous storms as well as a description of this type of storm.

CHAPTER 4. IF WEATHER AND CLIMATE ARE SIMPLE, WHY IS
PREDICTION SO HARD?

The U.S. Weather Service pages under "Verification" present many data sets related to the accuracy of forecasts. Again, kudos to this organization for addressing accuracy in such a data-intensive way.

A good discussion of forecast accuracy, and of the relative values of persistence and climatology in making forecasts, can be found on pp. 366–67 in

Meteorology Today, by Ahrens and Henson. In general, they say that persistence is only useful over a span of several hours, while climatology fades after ten days or so. They also say that forecasts are less accurate for precipitation than for temperature.

Detailed presentations on Chaos Theory and on Edward Lorenz can be found on Wikipedia.

Information on the environmental monitoring tower at the Harvard Forest can be found at the Harvard Forest site using those search terms.

Eddy covariance is explained on the Wikipedia site of that name. Figure 4.4 is from the Ameriflux Management Project located at the Berkeley Lab. Ameriflux is a national network of eddy covariance sites that share and compare methods and data. Fluxnet is a similar network at the global scale.

CHAPTER 5. EL NIÑO IS ONLY THE BEGINNING

Wikipedia has an extensive entry on El Niño that includes information on teleconnections. Searching for NOAA El Niño should bring you to a page geared more toward general readers. This page currently highlights several early workers on this topic before and after Gilbert Walker, who contributed to the concept.

Another Wikipedia page describes the Walker Circulation in some detail.

Entries for the California drought of 2013–2016 include a Wikipedia entry (actually entitled 2011–2017 at the moment). You can access annual rainfall data at the NOAA Climate at a Glance site. California has a series of excellent real-time pages on current water storage in reservoirs, and also current water content in the Sierra Nevada Mountains. Search under California Data Exchange Center.

The data on the Niño 3.4 index can be found on the U.S. Weather Service Climate Prediction Center page for El Niño/La Niña under Outlooks. There are also links there for the Arctic Oscillation (AO) and the North Atlantic Oscillation (NAO).

NOAA and other agencies use satellites, moored buoys, drifting buoys, sea level analysis, and expendable buoys to monitor conditions in the central Pacific Ocean related to El Niño/La Niña, as described on the web pages for this oscillation.

Wikipedia has a short entry on the Blob under the heading Ridiculously Resilient Ridge, but it might be more fun to seek out the cartoons picturing the battle between El Niño and the Blob.

The Wikipedia entry on the Polar Vortex is quite detailed. Cartoons can also be found! And yes, there is a Wikipedia page on Sudden Stratospheric Warming.

The Smithsonian article on Benjamin Franklin and the Gulf Stream by Kat Eschner appeared in May 2017. The figure for the actual Gulf Stream is drawn from a NOAA group of sites called SciJinks.

There is a Wikipedia page on the Atlantic meridional overturning circulation that includes comments on potential recent changes in the overall rate of flow of the Gulf Stream and velocity of AMOC. AMOC now is studied as part of the more general term "meridional overturning circulation" (MOC), which examines rates of flow in different parts of this global conveyor belt, as it has been termed; a single, connected system of surface and deepwater flows linking all the known major ocean currents, including the Walker circulation and the Gulf Stream. A joint U.S.-U.K. program called RAPID has undertaken a detailed monitoring effort for different parts of the global ocean circulation system. Descriptions of the methods used are available on the RAPID website.

AMOC and the Gulf Stream are one part of the global Thermohaline Circulation—a single, integrated cycle of water movement that links surface and deepwater currents, warm and cold currents, and alters climates throughout the Earth. Wikipedia has a detailed site by this title that includes a now classic graphic of the global direction of both surface and deep ocean currents.

CHAPTER 6. WHAT CAN ICE TELL US ABOUT RAPID CLIMATE CHANGE?

Wikipedia pages relating to this chapter include Ice Age, Louis Agassiz (this one is interesting as an example of a brilliant scientist who believed in some theories that were both wrong and racially biased), Milankovitch and the eponymous cycles, and pages on orbital eccentricity, axial tilt or obliquity, and precessions.

For a good visual representation of the changes in Earth's orbit and the impact in triggering ice ages, search the NASA site with the title: Milankovitch (Orbital) Cycles and Their Role in Earth's Climate.

At this writing, the *Physics Today* site has a good page on Tying Celestial Mechanics to Earth's ice ages.

The Wikipedia page on Ice Core has a good description of the coring device used and some of the processing involved, and it contains the graph used here on long-term changes in temperature and carbon dioxide concentration.

The Little Ice Age page on Wikipedia includes the fine-scale temperature record in figure 6.3, and there are pages for Heinrich and Dansgaard/Oeshger events, as well as Oxygen-18 and its uses in paleoclimatology. Figure 6.3 is modified from Wikipedia sites for global temperature change.

The eight-hundred-thousand-year-old ice core record for carbon dioxide is currently on the Climate.gov website but can be found reproduced in many places.

CHAPTER 7. EXTREMES WE WILL NEVER SEE

NASA maintains a site called Space Shuttle Image Gallery that contains a wealth of vibrant images of the Earth relayed from space.

The NASA website for Exoplanet Exploration includes a definition of the Goldilocks or Habitable zone. Wikipedia has a page entitled Circumstellar Habitable Zone.

Wikipedia pages available include Snowball Earth and Mikhail Budyko, while the pages for Mars and Venus include descriptions of the atmospheres of those two planets in the Goldilocks zone.

CHAPTER 8. HOW LONG DOES IT TAKE TO SHIFT A PARADIGM?

Michael Strevens's book presents a clear dichotomy between Popper and Kuhn, and between "normal" science and revolutions. A wealth of other presentations can also be found. The role of the Iron Rule has changed my way of thinking about how science is done, and I am constantly reminded of the analogy of the polyps building the coral reef. It is one of the best I have encountered.

For all that, the concept of "flow" in work is contrary to Strevens's view of the drudgery of normal science. Search "flow at work" or "flow in the workplace" for current references.

For sources on the Big Questions in Earth system science, I like Hallam's *Great Geological Controversies* for a concise and well-told set of science stories.

It is interesting that at the time of the book's first publication, in 1983, the question of the K/T extinction event was not completely settled, and that is apparent in the text.

Gould's defense of Ussher's method can best be found by a search for "Gould defends Ussher chronology." Wikipedia currently has an entry for this. On the larger question of the discovery of deep time, see Gould's *Time's Arrow, Time's Cycle*.

The statements by Hutton and Playfair as well as Darwin's sedimentation estimates can be found now on their Wikipedia pages. A search for "Darwin and age of Earth" will yield some good and some totally fatuous sites. William Thomson (Lord Kelvin) has a Wikipedia page that includes text and references on his approach to the age of the Earth.

An image of Hutton's Unconformity can be found on the Wikipedia page entitled Unconformity, or by a web search for "Hutton Unconformity drawing."

The coverage of the role of Marie Skłodowska-Curie (Madame Curie) in the discovery of radioactivity is less clearly presented on some sites. She stands alone among chemists and physicists with her Nobel Prize in each category, and her work stands out all the more given the difficulty women faced in their time in gaining support and recognition for their work. Her drive and accomplishments are captured in Barbara Goldsmith's *Obsessive Genius*.

The U.S. Geological Survey currently has an excellent page on using radiometric dating to determine the age of the Earth. For a concise technical presentation, I looked to Karl Turekian's *Global Environmental Change*.

Wegener and Taylor and the development of the theory of plate tectonics are well presented on Wikipedia. The entry for Continental Drift is more complete at this point in terms of history, while the entry for Plate Tectonics has more on the process. Any current map of the Earth or a globe will demonstrate what Wegener, and many others, noticed.

Wikipedia pages on the Mid-Atlantic Ridge, Magnetostratigraphy, and Paleomagnetism provide more details relevant to the theory of continents in motion. The NASA site on magnetic reversals is the source for the data on frequency of reversals. Data for the spreadsheet class exercise was adapted from J. A. Jacobs, *Reversals of the Earth's Magnetic Fields* (Cambridge: Cambridge University Press, 1994).

Nothing can beat Alvarez's telling of the story of the scientific revolution and paradigm shift represented by the K/T extinction and the process that led to the acceptance of a catastrophic explanation. Work continues, however, and so here are a few exceptions to the "no URLs rule."

The 2019 paper on the discovery of the endpoint of the tsunami surge in North Dakota is here: https://www.pnas.org/content/116/17/8190. Recent findings of iridium in the Chicxulub crater are here: https://advances .sciencemag.org/content/7/9/eabe3647?utm_campaign=toc_advances_2021- 02-26&et_rid=79631247&et_cid=3680819

A paper linking the meteor strike to the eruptions of the Deccan Traps is here: https://pubs.geoscienceworld.org/gsa/gsabulletin/article- abstract/127/11-12/1507/126064/Triggering-of-the-largest-Deccan-eruptions- by-the?redirectedFrom=fulltext

CHAPTER 9. THE STAGES OF ENVIRONMENTAL GRIEF

The psychological Five Stages of Grief are well summarized on the Wikipedia page of the same name.

The most recent book summarizing the work at Hubbard Brook is *Hubbard Brook: The Story of a Forest Ecosystem*, by Gene Likens and Richard Holmes.

U.S. EPA has an introductory site on Acid Rain and on the history of NAPAP. Wikipedia has pages on Acid Rain and the Acid Rain Program.

Wikipedia has a good site on Ozone Depletion. The NASA site called Ozone Watch contains both the latest data and the historical record of measurements of the ozone hole.

Wikipedia has a page on Malthus and an introduction to cornucopians. The quotation is from William Crookes and is drawn from an 1898 publication by the British Academy of Sciences.

Kean's book has an excellent summary of the history of the Haber-Bosch process and its inventors. For a history of the guano trade see Edward Melillo's work here: https://www.amherst.edu/system/files/MelilloAHR.pdf

Wikipedia has a good page on the dead zone phenomenon and includes references to the occurrence of such zones around the world. EPA has pages that can be found under Hypoxia Task Force.

CHAPTER 10. THE DISCOVERY (AND REDISCOVERY)
OF CLIMATE CHANGE

The retelling of the history of our understanding of climate change is based on the presentations in the three books cited as well as some information from Crawford's biography of Arrhenius. There are also Wikipedia pages for each of the major figures.

The quotation of journalism as a first rough draft of history has been attributed to writers in the editorial office of the *Washington Post* in the 1940s. A search using this phrase will reveal other versions of this story.

Information on the solar spectrum and the absorption of radiation in different regions of the spectrum can be found on many basic science websites. Wikipedia sites include: Prism, Photon, Electromagnetic Spectrum, and Greenhouse Gas. There is also a Wikipedia page on the Earth's Energy Balance.

The Keeling Curve, which should be the best-known environmental data set, can be found by searching with that term. At this point, the basic data are gathered on the Scripps Institution of Oceanography site hosted by UC San Diego.

Several good articles on the thirtieth anniversary of James Hansen's testimony to Congress that climate change was "real" are available. A web search could lead you to one by Elizabeth Kolbert in the *New Yorker* or newspaper articles in the *Guardian* and the *Boston Globe*.

The IPCC maintains a voluminous and dynamic website, including summaries and complete texts of the first through sixth reports. Currently that site is: https://www.ipcc.ch/. The title of the science report from the IPCC Sixth Assessment is: *AR6 Climate Change 2021: The Physical Science Basis*.

Weart's history has an in-depth discussion of the work of the IPCC as well as interesting discussions of the scientific and public response to the earliest report.

The ocean heat content data can currently be found on the Wikipedia page by that name. The weather.gov site has a story about Hurricane Katrina. Data on total Accumulated Cyclone Energy are from the Wikipedia page of the same name.

The Wikipedia page on Ocean Acidification is very complete.

Data on minimum sea ice extent in the Arctic Ocean is currently found on the NASA Global Climate Change page under Vital Signs—Arctic Sea Ice Minimum. That page currently has the dramatic video mentioned in the text. It's worth watching.

Data on global retreat or loss of ice from glaciers can be found on the National Snow and Ice Data Center page on Global Glacier Recession and the Climate Change: Mountain Glaciers page at Climate.gov.

Images of Status of Glaciers in Glacier National Park can be found on the U.S. Geological Survey site of the same name through a site named Climate Change Indicators: Glaciers. The site also has a link to the global data set on glacial retreat that is the source of the 68 percent reduction number.

Data on change in the mass of ice in Antarctica and Greenland are summarized at the EPA.gov site on the page Climate Change Indicators: Ice Sheets.

Wikipedia has an extensive and detailed page on Permafrost.

A story about the exploding sinkholes in Siberia is currently located here: https://www.bbc.com/future/article/20201130-climate-change-the-mystery-of-siberias-explosive-craters

The consecutive reports from the IPCC can be accessed from their primary website. Currently that site is: https://www.ipcc.ch/

CHAPTER 11. SIMPLIFYING THE EXPLANATION

As noted in the text, Arrhenius would not have created the equation used for the analysis reported in figures 11.1 and 11.2. The most recent mention of this equation was by Martin Walter (2010) in: http://www.ams.org/notices/201010/rtx101001278p.pdf. He assigns the term Greenhouse Law for CO_2 to the equation and credits Arrhenius for its formulation. A citation trail can be followed back from this entry, including a citation in the first IPCC report from 1990 (chap. 2, p. 51). Other early formulations of the effect of carbon dioxide on temperature also capture the nonlinear relationship used here, for example Hansen et al. (1988): https://agupubs.onlinelibrary.wiley.com/doi/abs/10.1029/JD093iD08p09341

Temperature data for figure 11.1 are from https://www.ncdc.noaa.gov/cag/global/time-series

The carbon dioxide data are from the Keeling Curve site, currently at: https://keelingcurve.ucsd.edu/

The Tyndall quote is currently on his page at NASA's Earth Observatory website.

More on Tyndall, Callendar, and Plass can be found on their Wikipedia pages.

As this book went into production, I was made aware of work by the American scientist Eunice Foote, who reported results of experiments on the effect of different gases on the retention of heat in vessels warmed by sunlight. She reported, in 1856, that carbon dioxide retained heat more effectively than others tested and concluded that higher concentrations of carbon dioxide in the atmosphere should increase Earth's surface temperature. This may be the first statement in the scientific literature linking carbon dioxide to global temperature. Tyndall described his methods in 1859 and reported his results on carbon dioxide and other gases in 1861. A recent paper with Scott Ollinger includes this update.

Eunice Foote, "Circumstances affecting the heat of the Sun's rays," *American Journal of Science and Arts, 22*, no. 66 (1856): 382–383, ia800802 .us.archive.org/4/items/mobot31753002152491/mobot31753002152491.pdf

John Tyndall, "Note on the transmission of radiant heat through gaseous bodies," *Proceedings of the Royal Society of London, 10* (1859): 37–39. https://www.jstor.org/stable/111604

John Tyndall, "I. The Bakerian Lecture.—On the absorption and radiation of heat by gases and vapours, and on the physical connexion of radiation, absorption, and conduction," *Philosophical Transactions of the Royal Society of London, 151* (1861). https://doi.org/10.1098/rstl.1861.0001

John Aber and Scott V. Ollinger, "Simpler presentations of climate change," *Eos, 103* (2022). https://doi.org/10.1029/2022EO220444

Additional estimates of climate sensitivity, including IPCC estimates, can be found on the Wikipedia page of the same name.

The 50 percent estimate for atmospheric retention can be found here: https://sos.noaa.gov/datasets/ocean-atmosphere-co2-exchange/

Predicted changes in this fraction with different levels of continued carbon dioxide emissions are presented in the "Summary for Policymakers" of the science document from the IPCC 6th Assessment.

The example cited for prediction of sea level rise compared with later measurements is from *The Copenhagen Diagnosis: Updating the World on the Latest Climate Science*, UNSW Climate Change Research Centre, UNSW Sydney NSW 205, figure 16, p. 37. https://oceanrep.geomar.de/id/eprint/11839/1/Copenhagen_Diagnosis_HIGH.pdf

The report on carbon dioxide and climate from the National Research Council of the U.S. National Academies can be found here: https://www.nap.edu/catalog/12181/carbon-dioxide-and-climate-a-scientific-assessment

The figure on changes in radiative forcing over time is from the EPA site on Climate Change Indicators: Climate Forcing.

Again, the IPCC reports are at: https://www.ipcc.ch/

CHAPTER 12. PREDICTIONS AND POSSIBLE SURPRISES

As this chapter deals with very recent projections of climate change impacts, actual URLs to cited websites are included. If those links become broken, searches on the major terms will, one hopes, bring you to the relevant information.

The science document from the Sixth IPCC Assessment is here: https://www.ipcc.ch/report/sixth-assessment-report-cycle/

Projections of the impact of current global commitments to reduce carbon dioxide emissions are in a recent BBC report quoting analysis by Climate Action Tracker. https://www.bbc.com/news/science-environment-59220687 https://climateactiontracker.org/global/temperatures/

Recent papers on Miocene and Eocene analogues to current climate are here: https://www.pnas.org/content/115/52/13288 https://www.researchgate.net/publication/307778046_A_suite_of_early_Eocene_55_Ma_climate_model_boundary_conditions

NASA estimates current mass loss of ice in Antarctica at 149 gigatons (10^{15} grams) per year. The Wikipedia page on Antarctic Ice Sheet gives the mass of ice there as 26,500,000 gigatons. Dividing the former by the latter gives an estimate of 177,852 years to complete disappearance. This assumes no change in the rate of loss. https://climate.nasa.gov/climate_resources/265/video-antarctic-ice-mass-loss-2002-2020/, https://en.wikipedia.org/wiki/Antarctic_ice_sheet

An analysis of the impacts of projected sea level rise on southern Florida, including the Everglades, by the National Park Service is here: https://www.nps.gov/articles/parkscience33-1_63-73_park_et_el_3860.htm

Recent papers on changes in AMOC velocity are here: https://www .nature.com/articles/s41561-021-00699-z/, https://www.nature.com/articles/ s41467-020-17761-w/

Permafrost loss and impacts are addressed here: https://arctic.noaa.gov/ Report-Card/Report-Card-2019/ArtMID/7916/ArticleID/844/Permafrost-and-the-Global-Carbon-Cycle

Data on global population, economic activity, and greenhouse gas emissions are from the World Bank: https://data.worldbank.org/about/get-started

Source for the 10 percent of greenhouse gases from agriculture for the United States is: https://www.epa.gov/ghgemissions/sources-greenhouse-gas-emissions and for one-third of global emissions is: https://www.nature .com/news/one-third-of-our-greenhouse-gas-emissions-come-from-agriculture-1.11708

Data on national wealth and consumption of meat, with a clear graphic, are here: https://ourworldindata.org/grapher/meat-consumption-vs-gdp-per-capita

Figure SPM.10 referenced in the text can be found on p. 41 of IPCC, 2021: "Summary for Policymakers." In *Climate Change 2021: The Physical Science Basis: Contribution of Working Group I to the Sixth Assessment Report of the Intergovernmental Panel on Climate Change*, ed. V. Masson-Delmotte, P. Zhai, A. Pirani, S. L. Connors, C. Péan, S. Berger, N. Caud, Y. Chen, L. Goldfarb, M. I. Gomis, M. Huang, K. Leitzell, E. Lonnoy, J. B. R. Matthews, T. K. Maycock, T. Waterfield, O. Yelekçi, R. Yu, and B. Zhou (Cambridge: Cambridge University Press, forthcoming). Currently available at: https://www.ipcc.ch/ report/ar6/wg1/downloads/report/IPCC_AR6_WGI_SPM_final.pdf

EPILOGUE

The Climate Solutions New England report for Southern New Hampshire is here: https://scholars.unh.edu/cgi/viewcontent.cgi?article=1002&co ntext=sustainability

The map of changing climate zones is from a report by the Union of Concerned Scientists entitled Confronting Climate Change in the U.S. Northeast. 2007. https://www.ucsusa.org/sites/default/files/2019-09/con-fronting-climate-change-in-the-u-s-northeast.pdf

Figure Credits

Figure Credits and the Creative Commons

There has been a very positive and recent change in the use and citing of figures and images. Most of the images accessed through Wikipedia, and increasingly in journals as well, are covered by the Creative Commons agreement, meaning that they can be freely copied and used with proper attribution to the source. I have tried to honor that approach here. In addition, the incredible wealth of information and images developed by government agencies in the United States is all in the public domain and has been used extensively in this book.

1.1–Originally published in *Zeitschrift für Physikalische Chemie,* volume 69, 1909. Photo engraving by Meisenbach Riffarth & Co. Leipzig. https://en.wikipedia.org/wiki/Svante_Arrhenius

2.2–Modified by the author from "The First Climate Model," NOAA, https://celebrating200years.noaa.gov/breakthroughs/climate_model/AtmosphericModelSchematic.png

3.1–"Seasons.svg," Wikimedia Commons, user PbBR8498, CC0 1.0. https://commons.wikimedia.org/wiki/File:Seasons.svg

3.2–"NASA depiction of earth global atmospheric circulation.jpg," Wikimedia Commons, modified by user Kaidor, CC BY-SA 3.0. https://en.wikipedia.org/wiki/Atmospheric_circulation#/media/File:Earth_Global_Circulation_-_en.svg

3.3–NOAA.gov, modified by U.S. National Weather Service. https://commons.wikimedia.org/wiki/File:Jetcrosssection.jpg

3.4–From "The Science behind the Polar Vortext," NOAA, updated March 2021. https://www.noaa.gov/multimedia/infographic/science-behind-polar-vortex-you-might-want-to-put-on-sweater

3.5–Created by the author.

3.6–NOAA. https://commons.wikimedia.org/wiki/File:Early_
January_2018_Nor%27easter_2018-01-04_1345Z.jpg

4.1–NOAA Weather Prediction Center. https://www.wpc.ncep.noaa.gov/
images/hpcvrf/maemaxyr.gif

4.2 (top)–Courtesy of David Babb, Department of Meteorology &
Atmospheric Sciences, The Pennsylvania State University. https://
www.e-education.psu.edu/files/meteo410/image/Lesson3/threat_
score0202.gif

4.2 (bottom)–NOAA. https://www.wpc.ncep.noaa.gov/images/hpcvrf/
wpc20yr.gif

4.3–Courtesy of William Munger, Harvard Forest. From "Environmental
Measurement Station Eddy Flux Tower," 2017. https://harvardforest
.fas.harvard.edu/photos/eddy-flux-tower

4.4–From "AmeriFlux Network and AmeriFlux Management Project,"
brochure, U.S. Department of Energy. https://ess.science.energy.gov/
ameriflux/

5.1–PAR. https://commons.wikimedia.org/wiki/File:LaNina.png

5.2–From "Can We Blame El Niño?" Climate.gov, October 23, 2009.
https://www.climate.gov/news-features/understanding-climate/can-
we-blame-el-ni%C3%B1o

5.3–From "ENSO: Recent Evolution, Current Status and Predictions,"
prepared by the Climate Prediction Center / NCEP for NOAA,
July 18, 2022. https://www.cpc.ncep.noaa.gov/products/analysis_
monitoring/lanina/enso_evolution-status-fcsts-web.pdf

5.4 (left)–"A Chart of the Gulf Stream," Library of Congress, Geography
and Map Division. Credit: World Digital Library. https://www.loc
.gov/resource/g9112g.ct000136/?r=0.106,0.062,0.926,0.603,0

5.4 (right)–From "What Is the Gulf Stream?" NOAA SciJinks. https://
scijinks.gov/gulf-stream/

6.1–Skeptical Science, CC BY-SA 4.0. https://skepticalscience.com/
graphics.php?g=342

6.2–Earth Observatory, NASA. https://earthobservatory.nasa.gov/features/
CarbonCycle/page4.php

6.3–Wikimedia Commons, user RCraig09, from Ed Hawkins, CC BY-SA 4.0. https://commons.wikimedia.org/wiki/File:2000%2B_year_global_temperature_including_Medieval_Warm_Period_and_Little_Ice_Age_-_Ed_Hawkins.svg

7.1 (top)–NASA. https://www.nasa.gov/sites/default/files/thumbnails/image/as17-148-22727_lrg_0.jpg

7.1 (bottom)–"Space Shuttle Discovery's Aft Cargo Bay," European Space Agency. https://www.esa.int/ESA_Multimedia/Images/2006/07/Space_Shuttle_Discovery_s_aft_cargo_bay

7.2–Wikimedia Commons, user Heinrich Holland, CC BY-SA 3.0. https://commons.wikimedia.org/wiki/File:Oxygenation-atm-2.svg

7.3–Wikimedia Commons, user Glen Fergus, CC BY-SA 3.0. https://commons.wikimedia.org/wiki/File:All_palaeotemps.svg

8.1–Map by U.S. Geological Survey. Reproduced from "Continental Drift," National Geographic Resource Library. https://www.nationalgeographic.org/encyclopedia/continental-drift/

8.2–From "Marine Geology and Geophysics, " NOAA, National Centers for Environmental Information. https://www.ngdc.noaa.gov/mgg/image/2minrelief.html

8.3–From Edward A. Mankinen and Carl M. Wentworth, "Preliminary Paleomagnetic Results from the Coyote Creek Outdoor Classroom Drill Hole, Santa Clara Valley, California," U.S. Geological Survey Open-File Report 03-187. https://pubs.usgs.gov/of/2003/ofo3-187/

8.4–NASA. https://en.wikipedia.org/wiki/File:Oceanic_spreading.png

9.1–From "Progress Report," U.S. Environmental Protection Agency. https://www3.epa.gov/airmarkets/progress/reports_2016-2017/acid_deposition_figures.html

9.2–From "World of Change: Antarctic Ozone Hole," NASA, Earth Observatory website. https://earthobservatory.nasa.gov/world-of-change/Ozone

9.3–From "Hypoxia," NOAA, National Ocean Service website. https://oceanservice.noaa.gov/hazards/hypoxia/

10.1–Wikimedia Commons, user Robert A. Rohde, CC BY-SA 3.0. https://commons.wikimedia.org/wiki/File:Solar_Spectrum.png

10.2–Created by the author, using data from "What Is Earth's Energy Budget? Five Questions with a Guy Who Knows," NASA. https://www.nasa.gov/feature/langley/what-is-earth-s-energy-budget-five-questions-with-a-guy-who-knows

10.3–Scripps CO_2 Program, Scripps Institution of Oceanography. https://scrippsco2.ucsd.edu/graphics_gallery/mauna_loa_record/mauna_loa_record.html

10.4–Created by the author, using data from NASA, "Arctic Sea Ice Minimum Extent." https://climate.nasa.gov/vital-signs/arctic-sea-ice/

10.5 (top)–From "Global Temperature Anomalies—Graphing Tool," NOAA. https://www.climate.gov/maps-data/dataset/global-temperature-anomalies-graphing-tool

10.5 (bottom)–From "Global Temperature Anomalies from 1880 to 2019," visualizations by Lori Perkins, January 15, 2020, Science Visualization Studio, NASA. https://svs.gsfc.nasa.gov/4787

11.1–Produced by the author, using data on carbon dioxide concentrations from the Keeling Curve at the Scripps Institution (https://scrippsco2.ucsd.edu/graphics_gallery/mauna_loa_record/mauna_loa_record.html) and global climate data from National Centers for Environmental Information, NOAA (https://www.ncdc.noaa.gov/cag/global/time-series/globe/land_ocean/ann/2/1880-2022).

11.2–Created by the author using published IPCC Fourth Assessment Data. https://www.ipcc-data.org/observ/ddc_co2.html

11.3–From "Climate Change Indicators: Climate Forcing," U.S. Environmental Protection Agency. https://www.epa.gov/climate-indicators/climate-change-indicators-climate-forcing

12.1–Created by the author, using data from the World Bank. https://data.worldbank.org/about/get-started

E.1–Adapted from Cameron Wake, Elizabeth Burakowski, Peter Wilkinson, Katharine Hayhoe, Anne Stoner, Chris Keeley, and Julie La Branche, "Climate Change in Southern New Hampshire," 2014, *The Sustainability Institute,* 2. https://scholars.unh.edu/sustainability/2. CC BY-NC 4.0.

E.2–Courtesy of the Union of Concerned Scientists (UCS). Adapted from
P. C. Frumhoff, J. J. McCarthy, J. M. Melillo, S. C. Moser, and
D. J. Wuebbles, "Confronting Climate Change in the U.S. Northeast:
Science, Impacts, and Solutions," Synthesis report of the Northeast
Climate Impacts Assessment (NECIA), Cambridge, Mass.: Union of
Concerned Scientists (UCS), 2007. https://www.ucsusa.org/sites/default/
files/2019-09/confronting-climate-change-in-the-u-s-northeast.pdf

Index

259

validation of models, 206–7
Venus, atmosphere of, 117, 118
visible light, 176, 179, 182–83, 185
volcanic activity: Arrhenius on, 16;
climate change and, 75; gases
released from, 117, 121, 123–24; ice
core dating and, 109–10; lava flows
from, 139, 149–50
Vostok ice core, 107*f*, 110

Walker, Gilbert, 80–82, 86, 102
Walker circulation, 82, 83*f*, 86
warfare, technology of, 166, 167
water vapor: cloud formation and, 45,
52, 58–59; condensation of, 45, 52,
56–58, 73, 117; evaporation of, 28,
45, 56, 73, 94, 183–84, 184*f*;
feedbacks involving, 207–8;
photosynthesis and, 71; radiation
absorption by, 14, 176, 180, 181*f*;
temperature and, 15, 45, 56, 193. *See
also* humidity
Weart, Spencer, 175, 177, 178, 193, 196,
201–2
weather: access to data on, ix–x;
climate compared to, 76; daily
patterns, 54–55, 55*f*; global patterns,
45–50; regional patterns, 50–54, 51*f*;

solar energy distribution and, 42.
See also atmosphere; meteorology;
oscillations; weather forecasting;
specific types of weather events
weather forecasting: chaotic systems
and, x, 31, 63, 68–74, 213; on
commercial sites, 54–55; models of,
25–30, 26*f*, 36; modern system of,
21, 31; now-casting, 68; wind
patterns and, 49–50. *See also*
accuracy of weather forecasts
Wegener, Alfred, 136, 137*f*, 142
westerly winds, 44*f*, 47, 48, 52
Wetherald, Richard, 208
Whole Earth images, 114, 115*f*
wildfires, 216
William of Occam, 20–21, 35. *See also*
Occam's razor
Winchester, Simon, 129
wind patterns: Arctic ice melt and,
189–90; complexity of, 46; Coriolis
effect and, 44*f*, 47–48, 81; at high
elevations, 49–50; Intertropical
Convergence Zone and, 44*f*, 81–82;
ocean currents and, 82; oscillations
and, 82–83; trade, 44*f*, 47, 48, 81,
82, 93–94; westerlies, 44*f*, 47,
48, 52